Lecture Notes in Computer Science 14545

Founding Editors

Gerhard Goos
Juris Hartmanis

The series Lecture Notes in Computer Science (LNCS), including its subseries Lecture Notes in Artificial Intelligence (LNAI) and Lecture Notes in Bioinformatics (LNBI), has established itself as a medium for the publication of new developments in computer science and information technology research, teaching, and education.

LNCS enjoys close cooperation with the computer science R & D community, the series counts many renowned academics among its volume editors and paper authors, and collaborates with prestigious societies. Its mission is to serve this international community by providing an invaluable service, mainly focused on the publication of conference and workshop proceedings and postproceedings. LNCS commenced publication in 1973.

Fabio Palomba · Carmine Gravino
Editors

Artificial Intelligence with and for Learning Sciences

Past, Present, and Future Horizons

First Workshop, WAILS 2024
Salerno, Italy, January 18–19, 2024
Proceedings

 Springer

Editors
Fabio Palomba
University of Salerno
Fisciano, Italy

Carmine Gravino 🆔
University of Salerno
Fisciano, Italy

ISSN 0302-9743   ISSN 1611-3349 (electronic)
Lecture Notes in Computer Science
ISBN 978-3-031-57401-6   ISBN 978-3-031-57402-3 (eBook)
https://doi.org/10.1007/978-3-031-57402-3

This Springer imprint is published by the registered company Springer Nature Switzerland AG
The registered company address is: Gewerbestrasse 11, 6330 Cham, Switzerland

If disposing of this product, please recycle the paper.

Preface

We cordially welcome you to the proceedings of the 1st Workshop on *Artificial Intelligence with and for Learning Sciences: Past, Present, and Future Horizons* (WAILS 2024), held on January 18–19, 2024, in Salerno, Italy.

WAILS 2024 provided a quality forum for researchers and practitioners from academia, industry, and government to present and discuss previous achievements, current challenges and solutions, and future perspectives concerned with the adoption of artificial intelligence methods and techniques in the context of learning sciences.

We received twenty-four submissions, which underwent a thorough single-blind review with each submission receiving feedback from at least four expert reviewers. Following extensive discussion amongst the reviewers and chairs, fourteen submissions were accepted as full papers and five as extended abstracts.

The contributions dealt with several interesting aspects and the program was organized into three technical sessions:

– Metaverse and Virtual Reality
– Inclusion and Gamification
– Data-driven Technologies for Education

A discussion session, a panel entitled "Toward Educational Metaverse: How Far Are We?" and a keynote speech completed the program of WAILS 2024.

The keynote speech was on "Ethical Personalization in Education through Responsible Artificial Intelligence" and was given by Mirko Marras, who is a Researcher from the University of Cagliari, Department of Mathematics and Informatics.

The approved submissions, along with the keynote address, form an engaging agenda that offers numerous fresh ideas and perspectives, while also proposing potential avenues for future research.

Our gratitude extends to all the authors who submitted papers, as well as to the members of the program committee for their substantial and constructive feedback to the authors. Additionally, we express our appreciation to the organizing committee of WAILS 2024 for their assistance in managing the logistical aspects of the event.

Fabio Palomba
Carmine Gravino

Organization

General Chair

Sibilio, Maurizio University of Salerno, Italy

Program Committee Chairs

Gravino, Carmine University of Salerno, Italy
Palomba, Fabio University of Salerno, Italy

Local Arrangement Chairs

Di Tore, Stefano University of Salerno, Italy
Sellitto, Giulia University of Salerno, Italy

Proceedings Chairs

Barra, Paola University of Naples Parthenope, Italy
Pecorelli, Fabiano University of Salerno, Italy

Web Chair

Di Dario, Dario University of Salerno, Italy

Social Media Chair

Bilotti, Umberto University of Salerno, Italy

Program Committee Members

Aiello, Paola	University of Salerno, Italy
Bouwmans, Thierry	University of La Rochelle, France
Colazzo, Salvatore	University of Salerno, Italy
Di Pace, Anna	University of Foggia, Italy
Di Tore, Alfredo	University of Cassino, Italy
Di Tore, Stefano	University of Salerno, Italy
Dimitriadis, Yannis	University of Valladolid, Spain
Eleni, Ilkou	Leibniz University Hannover, Germany
Ferrucci, Filomena	University of Salerno, Italy
Fornasari, Alberto	University of Bari, Italy
Freire Obregón, David Sebastián	University of Las Palmas de Gran Canaria, Spain
Hlosta, Martin	Swiss Distance University of Applied Science, Switzerland
Lazzari, Marco	University of Bergamo, Italy
Marras, Mirko	University of Cagliari, Italy
Merlini, Donatella	University of Firenze, Italy
Moreno Guerrero, Antonio José	University of Granada, Spain
Rida, Imad	University of Technology of Compiègne, France
Tharewal, Sumegh	MIT World Peace University, India
Todino, Michele Domenico	University of Salerno, Italy
Viola, Ilaria	University of Salerno, Italy
Zappalà, Emanuela	University of Salerno, Italy

Extended Abstracts

Understanding Practices, Needs and Challenges of Lifelong Learning Digitalization

Francesco Beccuti[1], Gianni Fenu[1], Francesca Maridina Malloci[1], Mirko Marras[1],
Antonello Mura[2], Silvio Marcello Pagliara[2], and Maria Polo[1]

[1] Department of Mathematics and Computer Science, University of Cagliari, Via Ospedale 72, 09124 Cagliari, Italy
[2] Department of Literature, Languages and Cultural Heritage, University of Cagliari, Via Is Mirrionis 1, 09123 Cagliari, Italy

Abstract. Delving on a mixed-method approach, we report on a preliminary exploration concerning the attitude towards digitalization of organizations involved in lifelong learning. Our findings point to a positive inclination towards digital transformation motivated by wishes to improve effectiveness, personalization and monitoring. However, organizations report obstacles connected to limited hardware and software infrastructure as well as educational and technological know-how.

Keywords: Digitalization · Education Technologies · Lifelong Learning

1 Introduction

Modern information technologies bring about both opportunities and challenges with broad implications for society. European, national and regional regulations tend to emphasize the role of these technologies, especially those involving artificial intelligence, in connection with the desire to enhance well-being and social welfare. The link between digitalization, well-being and welfare would need to be critically examined in relation to the political and educational dimension. Hence, the aforementioned regulations often advocate for initiatives aimed at better understanding the multifaceted phenomenon of digitalization. One of these initiatives, the project "Ecosystem of Innovation for Next Generation Sardinia", involves investigating digitalization within training organizations (operating in the region of Sardinia, Italy) also with reference to the adoption of data-driven intelligence. Indeed, lifelong learning is a field where digitalization is growing, but training organizations may experience difficulties in engaging with this process.

Working within the project, we thus started to explore the practices, needs and challenges reported by local training organizations, by employing a mixed-method approach based on surveys and interviews. Preliminary results show that these organizations display openness to digitalization, especially in view of their desire to offer more personalized learning, enhance learning process monitoring, implement effective digital teaching models as well as ameliorate learning process quality evaluation and monitoring. However, these organizations often report to lack the technological infrastructure as well as the necessary know-how to manage the complexity involved in their implementation.

2 Method

To gather information about the practices, needs and challenges related to these organizations' functioning, we relied on the following mixed-method approach.

Step 1: Structured surveys. We distributed a survey to training organizations involved in lifelong learning (e.g., professional associations, public administrations) to gather information on typical learning context and practices, including questions on learner profiles (e.g., qualifications, interests), training (e.g., courses information, design, delivery) and adopted technologies (e.g., tools, usage).

Step 2: Semi-structured interviews. We organized focused meetings aimed to reach a deeper understanding of the content analyzed in the aforementioned surveys. The interviews employed questions on the training activity design and delivering as well as on the ways technologies are employed in relation to these, with an emphasis on the challenges faced by the organizations.

Step 3: Inception workshops. We organized a series of inception workshops aimed to provide insight for designing supportive data-driven technology which could be of interest to the selected organizations. This involved personas co-creation, user scenarios definition and requirements elicitation for pilot use cases.

3 Preliminary Findings

Survey results portray a diverse educational landscape both in terms of the organizational affiliations of the learners and in terms of their academic qualifications. In-person training activities (including face-to-face lectures, individual/group activities and laboratory activities) appear predominant in the surveyed organizations' functioning. Furthermore, the assessment methods employed by the organizations encompass both formative and summative evaluations, with a prevalence of closed-ended questionnaires. Moreover, we found that technologies are mostly used for content creation, quality monitoring and web video conferencing. In-depth interviews showed a growing interest of the surveyed organizations for adopting technology-based solutions (in order to convey, in particular, facilitated course management, scalable course delivering and artificial-intelligence-based support) and also highlighted possible challenges, primarily related to a lack of know-how and adequate infrastructure. During the workshops, we explored possible ways to address these challenges by focusing on four prototypes aimed at attending to learning process monitoring, general monitoring, learning personalization and quality management. The next steps in the project involve consolidating these, assessing their effectiveness and problematizing their impact both theoretically as well as via laboratory studies and user-related studies.

Acknowledgment. We acknowledge financial support under the National Recovery and Resilience Plan (NRRP), Miss. 4 Comp. 2 Inv. 1.5 - Call for tender No.3277 published on Dec 30, 2021 by the Italian Ministry of University and Research (MUR) funded by the European Union – NextGenerationEU. Prj. Code ECS0000038 eINS Ecosystem of Innovation for Next Generation Sardinia, CUP F53C22000430001, Grant Assignment Decree N. 1056, Jun 23, 2022 by the MUR.

Generative Artificial Intelligence to Improve the Learning of Test-Driven Development

Pietro Cassieri, Simone Romano⑩, and Giuseppe Scanniello⑩

University of Salerno, Salerno, Italy
{pcassieri,siromano,gscanniello}@unisa.it

Abstract. *Test-Driven Development* (*TDD*) can be beneficial not only in "traditional" software development contexts but also in other development contexts such as in the implementation of *Embedded Systems* (*ESs*). However, learning TDD, as well as correctly applying it, is not easy, especially in the implementation of ESs. To ease the learning process of TDD in such a context, we envisage using generative *Artificial Intelligence* (*AI*).

Keywords: TDD · AI · Embedded System

1 Background and Motivation

Test-Driven Development (*TDD*) is an agile software development practice [2] that promotes short cycles, composed of three phases each, to incrementally implement software functionality:

Red Phase - *write a unit test for a small chunk of functionality not yet implemented and watch the newly written test fail*;
Green Phase - *make the newly written test pass as quickly as possible, committing whatever "sin" is necessary to do so, and watch all unit tests pass*;
Refactor Phase - *refactor the code, thus remedying any sin previously committed, and watch all unit tests pass.*

A TDD practitioner repeats the Red-Green-Refactor cycle as long as the tackled functionality is not entirely implemented; when this happens, s/he is allowed to tackle new functionality. TDD is claimed to improve functional and internal quality, as well as developers' productivity [6]. Nevertheless, learning TDD as well as correctly applying it is not easy. For example, TDD newcomers (students or developers learning TDD) often *(i)* find counter-intuitive the *test-first dynamic* behind TDD (*i.e.,* writing a test before the corresponding production code), thus perceiving TDD difficult; *(ii)* skip the Refactor phase despite being refactoring opportunities [5], and *(iii)* underestimate the importance of the Red phase [5]. Greening [4] claims that TDD would be beneficial also in the *Embedded System* (*ES*) development context. Although developing ESs comes with a unique set of challenges (*e.g.,* the underlying hardware) in addition to the challenges faced when developing traditional software [4], Esposito *et al.* [3] recently showed that the use of TDD in the ES development context is promising (*e.g.,* they observed that TDD improved the functional quality of ESs).

2 General Aim and Proposed Innovation

Our research aims to ease the learning process of TDD in the ES development context. In particular, we are going to provide TDD newcomers with an Artificial Intelligence (AI) code assistant, based on *ChatGPT*, plugged into their Integrated Development Environments (IDEs). The code assistant leaves a TDD newcomer the duty of writing a failing unit test in the Red Phase. The code assistant intervenes only during the Green Phase in which it writes the production code, in place of the TDD newcomer, to make the unit tests pass. Finally, it is up to the TDD newcomer to refactor the code in the Refactor phase. We want to push TDD newcomers to focus much more on two phases, Red and Refactor, which are often deemed less important or even skipped by TDD newcomers [5].

3 Early Achievements and Future Research Direction

To understand the feasibility of our proposal, we implemented an early prototype [1] of our AI code assistant tool. We preliminary tested this prototype on a series of ES implementation tasks assigned to the students enrolled in the ES course in the a.y. 2023–2024. We observed that, in most cases, our code assistant makes the unit tests pass in the Green Phase. In other cases, the code assistant still provides production code but does not pass all unit tests, thus requiring human intervention. This is not a big issue since even human beings are something not able to make all unit tests pass on the first try. Finally, we observed that the production code provided by our code assistant becomes progressively more complex, thus encouraging developers to refactor it. From a pedagogical standpoint, this also transfers ownership of the production code to the developer.

These early achievements encouraged us to experiment with our code assistant on the field. In particular, we are going to introduce our code assistant in the next edition of the *Embedded Systems* course (held at the University of Salerno in the a.y. 2024–2025)—in which students are asked, among other things, to learn to use TDD when developing ESs—and then assess whether the code assistant eases the learning process of TDD by newcomers.

References

1. Gai4-tdd repository. https://github.com/Hauntlight/GAI4-TDD
2. Beck, K.: Test-Driven Development: By Example. Addison-Wesley, Boston (2003)

3. Esposito, M., Romano, S., Scanniello, G.: Test-driven development and embedded systems: an exploratory investigation. In: Proceedings of the SEAA, pp. 239–246. IEEE (2023)
4. Grenning, J.: Test Driven Development for Embedded C. Pragmatic Bookshelf (2011)
5. Romano, S., Fucci, D., Scanniello, G., Turhan, B., Juristo, N.: Results from an ethnographically-informed study in the context of test driven development. In: Proceedings of the EASE, pp. 10:1–10:10. ACM (2016)
6. Romano, S., Zampetti, F., Baldassarre, M.T., Di Penta, M., Scanniello, G.: Do static analysis tools affect software quality when using test-driven development? In: Proceedings of the ESEM, pp. 80–91. ACM (2022)

Toward the Deployment of a Chatbot to Augment Computer Science Education

Giusy Annunziata, Giulia Sellitto, Stefano Lambiase, Emanuele Bruno,
Gabriele De Vito, and Filomena Ferrucci

University of Salerno, Italy

Extended Abstract

Over the last few years, human life has been significantly improved by many techno-logical innovations, especially due to considerable advances in the field of Artificial Intelligence. In particular, the most noticeable impact of such progress on everyday life is given by *Conversational Agents*, also known as *Chatterbots*, and later shortened to *Chatbots* [2]. Such solutions consist of computer applications capable of simulating conversations with one or more human users in a lifelike way; in certain cases, chatbots make themselves perceived as so realistic that humans do not realize they are in fact talking with a machine. Currently, the use of chatbots is increasingly frequent in many areas, such as (1) customer service, in which chatbots assist users in finding quick and easy solutions to most common problems, (2) job assistance, provided to practitioners working on various tasks, and (3) teaching support in academic contexts, where chatbots accompany the teacher, explaining in a more interactive way some topics that may be complex for students [1, 5].

Chatbots in education are becoming more and more popular due to the advantages they guarantee toward personalized learning, ease of use, and accessibility. Such bene-fits have been observed in various contexts, from primary and secondary education to university and vocational training [3]. One of the most recent chatbots proposed in the literature is *Hermias*, presented by Petousi et al. [4], which is aimed at helping high school students in the learning of History. The chatbot falls under the *Bots of Conviction (BoCs)*, which shift the focus from offering information to provoking reflection. Namely, Hermias seeks to encourage contemplation among students by engaging in conversation with them, assisting in the study of History, and encouraging historical empathy through the persona of a young slave in the Ancient Agora of Athens, Greece.

We aim to follow the path traced by Petousi et al. [4], by developing a chatbot with three-fold intentions, i.e., we are interested in (1) encouraging the engagement of high school students in the field of Computer Science, particularly about the topic related to networks, (2) improving their interest toward the history of Computer Science and the progress of this discipline over the decades, and (3) raising students' awareness on women in Science, escaping the common belief that *STEM* subjects are mainly for males.

Therefore, we propose *CAIHL*, a chatbot envisioned to help students understand the concepts of computer networks, with an eye on the history of the subject and its pioneers. In particular, CAIHL impersonates *Hedy Lamarr*, a leading figure in Computer Science who made substantial contributions to the field of networks. Known both as a scientist and actress, she is widely recognized as the matriarch of the technologies underpinning modern Wi-Fi. As such, she is an ideal candidate for elucidating fundamental concepts of computer networks and wireless transmission protocols to students, and also encouraging female students to pursue their interests in STEM fields. Through the persona of Hedy Lamarr, students can be captured in the learning of computer science, as the curiosity in her personal history acts as a Trojan horse, which initially engages students by means of history and trivia, and finally brings them to learn. While Hermias [4] represents a historical figure whose main purpose is to encourage the study of History itself, CAIHL is aimed at tickling students' curiosity via the presentation of the fascinating persona of a scientist-and-actress who really existed, to capture their interest and then shift it toward computer science.

We plan to evaluate CAIHL in a real context, by experimenting with its usage in a high school classroom setting. Namely, we design our investigation based on the pedagogical method of participatory learning, which shifts the focal point away from the teacher and toward the students themselves. The primary objective is to foster communication and integration among students, empowering individuals to actively participate and engage with both the teacher and their peers. In such a context, the teacher's role evolves to a facilitator, guiding discussions, posing questions, and coordinating contributions, while the students actively engage through the means of the chatbot. We hypothesize that the employment of CAIHL in a real education environment can boost the effectiveness of lectures, by stimulating the curiosity of students and making them the protagonists of the learning.

References

1. Brandtzaeg, P.B., Følstad, A.: Why people use chatbots. In: Kompatsiaris, I., et al. (eds.) Internet Science. INSCI 2017. LNCS, vol. 10673, pp. 377–392. Springer, Cham (2017). https://doi.org/10.1007/978-3-319-70284-1_30
2. Deryugina, O.: Chatterbot. Sci. Tech. Inf. Proc. **37**, 143–147 (2010). https://doi.org/10.3103/S0147688210020097
3. Kuhail, M.A., Alturki, N., Alramlawi, S., Alhejori, K.: Interacting with educational chatbots: a systematic review **28**(1) (2022). https://doi.org/10.1007/s10639-022-111 77-3
4. Petousi, D., Katifori, A., McKinney, S., Perry, S., Roussou, M., Ioannidis, Y.E.: Social bots of conviction as dialogue facilitators for history education: promoting historical empathy in teens through dialogue. In: Interaction Design and Children (2021)
5. Smutny, P., Schreiberova, P.: Chatbots for learning: a review of educational chatbots for the Facebook messenger. Comput. Educ. **151**, 103862 (2020)

BitStory—An Immersive Serious Game for Computer Science Education

Giusy Annunziata, Raffaele Aurucci, and Filomena Ferrucci

University of Salerno, Salerno, Italy

Keywords: Serious Game · Education · Computer Science

Introduction Introducing computer science education in primary school is important in today's digital age since it can provide students with a solid foundation for problem-solving, critical thinking, and computational skills [4]. This motivated many initiatives in the world targeting K-12 computer science education, such as Computer Science for All (CSforAll)[1] and Code Week. However, traditional teaching methods often are ineffective in stimulating students' interest, so innovative approaches are necessary.

Gamification [5] is an emerging approach aiming to address the limitations of traditional teaching methods by incorporating game features, such as challenges, rewards, and competition, into educational environment, to make learning more interactive and fun. *Serious Games* [1] are game created with a specific educational goal which combine gamification and the simulation of a real context in a protective and immersive environment to motivate and stimulate students, harnessing their natural curiosity and enthusiasm, and making education both effective and enjoyable [7].

Dehghanzadeh et al. [3] investigated the impact of gamification in K-12, highlighting it can be an effective learning strategy that can motivate and engage students. Integrating such games into educational environments brings notable advantages. Our objective is to enhance the relevance of these benefits in the realm of a highly impactful subject today—Computer Science.

We aim to create *BitStory*, a serious game geared towards primary school children, fostering an interest in computer science. Players navigate an avatar through a world populated by computer science pioneers, engaging in interactive conversations to learn about their contributions. Each character presents a topic through an interactive mini-game, employing artificial intelligence to adapt complexity levels and provide constructive feedback, enhancing the educational experience.

Game Description BitStory is located in a large and timeless room, represented by a huge desk with different Computers—illustrated in Fig. 1. When the player walks near an avatar representing a well-known figures in the Computer Science field, the game will ask the player to start a conversation with that character and move them to a scene of the historical age of that character. The game will show the player a list of buttons according to a specific topic (life of the character, main contribution to computer science, other contributions). The player will choose a topic, and the avatar will begin its

---

[1] CSforAll https://www.csforall.org/about/.

explanation; after, the avatar will repeat the player to choose possible topics to continue the conversation.

Ada Lovelace, Alan Turing, and Hedy Lamarr are examples of characters present in BitStory [6]. Each will also offer an interactive game related to a specific Computer Science topic. For example, after the conversation with Alan Turing, who will tell about his life and how he deciphered "Enigma" in World War II, the game will ask the player to play a similar mini-game. The player will decrypt a message using a cipher, in which each letter of the alphabet is associated with another letter, number, or special character. The game will be provided with an artificial intelligence component, which will suggest improvements or explain to students when they make mistakes. The AI will also automatically generate exercises and the cipher, based on students' ages and abilities. In this way, the game will continue to engage and stimulate students.

An example of how the game will appear is shown in Fig. 1.

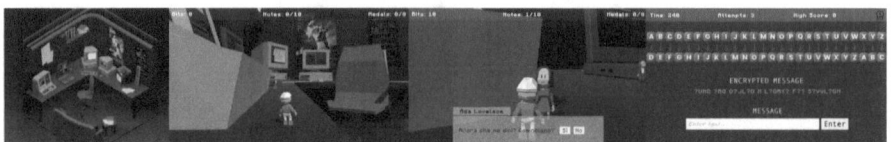

Fig. 1. The player moves through the world of BitStory and interacts with characters.

Preliminary Results and Future Work The current version of BitStory is only a prototype, representing an early stage of development—the prototype code is available [2]. We aim to use it to improve the game using an exploratory approach involving end users (students) in identifying the most effective features.

We intend to improve BitStory, integrating further information among characters to improve conversations and implementing all the mini-games to improve the educational and engaging factors. We also aim to improve the graphics, realizing avatars more similar to the historical characters and context they represent. Once BitStory will be ultimate, we intend to assess it with K-12 students in a controlled experiment through a case study. We will take advantage of a questionnaire regarding comprehension of topics. We will compare the questionnaire results from a control group and a group that will be subjected to using BitStory to evaluate it qualitatively.

References

1. Ahrens, D.: Serious games–a new perspective on workbased learning. Procedia-Soc. Behav. Sci. **204**, 277–281 (2015)
2. Annunziata, G., Aurucci, R., Ferrucci, F.: Bitstory - online appendix (2023). https://github.com/raffaele-aurucci/BitStory
3. Dehghanzadeh, H., Farrokhnia, M., Dehghanzadeh, H., Taghipour, K., Noroozi, O.: Using gamification to support learning in K-12 education: a systematic literature review. Br. J. Educ. Tech. https://doi.org/https://doi.org/10.1111/bjet.13335

4. Garneli, V., Giannakos, M.N., Chorianopoulos, K.: Computing education in K-12 schools: a review of the literature. In: 2015 IEEE Global Engineering Education Conference (EDUCON), pp. 543–551 (2015). https://doi.org/10.1109/EDUCON.2015.7096023
5. Kiryakova, G., Angelova, N., Yordanova, L.: Gamification in education. In: Proceedings of 9th International Balkan Education and Science Conference (2014)
6. O'Regan, G.: Giants of computing. A Compendium of Select, Pivotal Pioneers. Springer, London (2013). https://doi.org/10.1007/978-1-4471-5340-5
7. Ullah, M., et al.: Serious games in science education. a systematic literature review. Virtual Reality Intell. Hardware **4**(3), 189–209 (2022)

ProjectCoding—Applied Game for Learning Java

Giusy Annunziata, Stefano Lambiase, Alessio Piro, Dario Di Nucci,
and Filomena Ferrucci

University of Salerno, Salerno, Italy

Keywords: Gamification • Programming Languages • Education

Introduction. In today's digital age, software plays a key role in almost every aspect of our lives. Acquiring proficiency in programming languages, e.g., Java, is necessary due to business demand. Nevertheless, identifying ways to engage and motivate students in learning activities is a significant educational issue [3]. Appropriate educational methods are critical to mastering the tools that drive modern technology and innovation, enabling individuals to shape the future and solve complex problems in our rapidly changing digital world.

Gamification adopts elements belonging to the world of games in learning contexts [3]. This technique has become extremely popular over the years and has slowly given rise to using video games with specific educational goals other than pure entertainment. Among the most recent techniques, *Applied Games* are video games developed to educate players [5]. While the primary goal of gamification is to add game elements within a real-world context, Applied Games aim to insert educational components within a game-oriented environment [4].

In literature, Chang et al. [2] developed a game to engage students when exercising and assessing their competencies in Java. The game motivates and challenges the students while ensuring task focus. It is compatible across various operating systems and utilizes AJAX for rapid responses and immediate interplayer interactions.

In this paper, we contribute to the sets of tools that support learning Java, one of the most used programming languages in the academic and practitioner fields, by developing *ProjectCoding*, an Applied Game focusing on learning the Java programming language and equipped with an Artificial Intelligence component to generate game paths considering the players' skill levels randomly.

Game Design. To develop ProjectCoding, we relied on Godot Engine[2], an open-source tool specifically for making video games as mobile applications. ProjectCoding is a "roguelite game" in which the player must overcome a series of challenges, gaining experience and acquiring resources needed to tackle subsequent challenges. The game map, illustrated among the scenarios in Fig. 1, is designed as a tree where each node is a stage: a Battle, an Event, a Luna's Shop, a Treasure, and a Boss. The player's goal is to get to the bottom of the tree and defeat the boss at the end of the level. Among the different stages, there is Luna's Shop, which allows the player to purchase code snippets,

---

[2] Godot Engine https://godotengine.org.

as shown in Fig. 1. Treasure is a stage where players will be rewarded with experience or exclusive gadgets to upgrade their character. Event stage represents an event that confronts the player with a choice that can lead to positive or negative consequences. Once players have acquired resources and code snippets, they can combine them to construct a code that executes their plan of attack in battles. Arriving at the Battle stage, the player will face an opponent using the attack plan developed earlier to join the code snippets in his possession, as shown in Fig. 1. When choosing a stage, the player will advance through the map until the final Boss. ProjectCoding differs from other games already in the literature. It is geared toward mobile platforms to be constantly accessible online [1]. It provides game resources and interactive materials and is an applied game composed of more engaging game elements.

Fig. 1. Scenarios: Luna's Shop, Procurement, Treasure, Game Map, Code Creation

Future Implementation. Since ProjectCoding is a prototype, we developed only one boss and a few enemies in stages featuring few resources. As the first future implementation, we want to introduce new enemies with different features, elements, and attacks. Then, we aim to consider new types of resources, considering the different elements of the programming languages. Last, we want to improve the bosses by creating a different combat system with dynamic coding development. Given that only one map exists, we aim to integrate an artificial intelligence algorithm to generate maps based on the player's experience and resources, balancing the stages to be entered. Once the listed improvements are built, we intend to conduct a controlled experiment to validate the applied game and test its effects on students.

References

1. Annunziata, G., Lambiase, S., Piro, A., Ferrucci, F.: Projectcoding—applied game for learning Java (2023). https://github.com/AlessioPiro/ProjectCodingDemo
2. Chang, M., Kinshuk: web-based multiplayer online role playing game (MORPG) for assessing students' Java programming knowledge and skills. In: 2010 Third IEEE International Conference on Digital Game and Intelligent Toy Enhanced Learning, pp. 103–107 (2010). https://doi.org/10.1109/DIGITEL.2010.20
3. Kiryakova, G., Angelova, N., Yordanova, L.: Gamification in education. In: Proceedings of 9th International Balkan Education and Science Conference, vol. 1, pp. 679–684 (2014)

4. Laine, T.H., Lindberg, R.S.: Designing engaging games for education: a systematic literature review on game motivators and design principles. IEEE Trans. Learn. Technol. **13**(4), 804–821 (2020)
5. Plass, J.L., Homer, B.D., Kinzer, C.K.: Foundations of game-based learning. Educ. Psychol. **50**(4), 258–283 (2015)

Contents

Designing a Collaborative Safety Training Experience in Virtual Reality

Paola Barra[2], Andrea Antonio Cantone[1]([⊠]), Rita Francese[1],
Marco Giammetti[1], Raffaele Sais[1], Otino Pio Santosuosso[1],
Aurelio Sepe[1], Simone Spera[1], Genoveffa Tortora[1],
and Giuliana Vitiello[1]

[1] Department of Computer Science, University of Salerno, 84084 Fisciano, SA, Italy
{acantone,francese,tortora,gvitiello}@unisa.it,
{r.sais,o.santosuosso,a.sepe21,s.spera7}@studenti.unisa.it
[2] University of Parthenope, Napoli, Italy
paola.barra@uniparthenope.it

Abstract. Recently, the utilization of videoconferencing applications has had a substantial increase in a wide range of domains, including meetings, conferences, educational activities, and safety training. The emergence of the COVID-19 pandemic has further highlighted the pivotal role these tools play in how businesses and public institutions coordinate and oversee their employees' activities. They offer the distinct advantage of accommodating a virtually unlimited number of participants, surmounting geographical constraints, and curbing the costs associated with organization and travel, all while contributing to improved environmental sustainability.

Nevertheless, the challenge of maintaining social interaction and engagement remains when individuals work remotely using such tools, particularly in scenarios where physical proximity with others is necessary for effective learning or training processes.

Safety training is a critical aspect of disaster preparedness and prevention. Traditional safety training methods often lack realism and fail to adequately prepare people for real-world emergencies, especially in a cooperative context.

This article presents the design of an immersive multi-user experience of safety training and proposes a collaborative environment for fire extinguisher training by using the Meta Quest 2 headset. Learning the procedures before operating in a real setting is the way to avoid operators' mistakes can have serious consequences on process operation and safety. We also described how to organize the training activities and conduct the evaluation.

Keywords: virtual reality · cooperative environment · safety training

© The Author(s), under exclusive license to Springer Nature Switzerland AG 2024
F. Palomba and C. Gravino (Eds.): WAILS 2024, LNCS 14545, pp. 1–10, 2024.
https://doi.org/10.1007/978-3-031-57402-3_1

1   Introduction

Due to COVID-19 pandemic video conferencing platforms, such as Google Meet [16], Microsoft Teams [22], and Zoom [9] have seen a significant increase in daily participants since March 2020. However, a drawback of those online platforms is the lack of eye contact and body language, which cannot always be transmitted via video conferencing, and the lack of shared space [14].

Cooperative virtual environments have made possible greater social interaction between users that was not possible with video conferencing platforms. In a 3D virtual environment, called Metaverse, it is possible to replicate the real world and to interact socially in space using avatars. Examples of precursors of this interactive virtual space at the beginning of two thousand years were online virtual worlds, such as Second Life [6], The Sims, Fortnite, World of Warcraft, and Minecraft [1]. Many relevant organizations, ranging from IBM to N.A.S.A., owned their Second Life Island, which was also used to support collaboration.

Currently, the technologies to access this virtual Metaverse adopted Virtual Reality (VR) and Augmented Reality (AR) are cheaper than in the past, when those models were too expensive and invasive to be widely used [11,23]. There are several fields of applications where the Metaverse adoption is growing significantly: mainly in gaming[1] but also in medical [15], sports [17], education [5,12] domains and safety training [10].

To prevent accidents during intervention risky activities, it is important to train the workforce to follow the correct procedures and act fast in emergency situations [10]. Commonly utilized training approaches encompass a PowerPoint lecture, computerized simulations, online learning modules, familiarization with safety and/or production materials, and hands-on practice in either simulated or actual production environments. The latter usually necessitates the presence of an in-person supervisor who offers direction and identifies errors throughout the training phase. Recently, immersive solutions have been experimented with, such as the adoption of this technology with new VR devices.

In this paper, we present the design of a training experience for the use of fire extinguishers carried out using the multiuser platform MetaCUX [2,3], an immersive virtual environment in which it is possible to carry out training, meetings and explore multi-user virtual experiences cooperatively and collaboratively. A new scenario has been designed to support fire extinguishing training together with the training procedure and the associated assessment.

2   Background

Recently, safety has been one of the main concerns in the construction industry, and when conducting risky tasks in general. To achieve the target of mitigating incidents there is a need to perform better education and training activities for workers who perform this kind of task [19].

---

[1] https://www.warnerbros.it/gioco/genere-avventura/batman-arkham-vr/.

In the Metaverse people are totally involved and the distance among them is deleted [7]. Researchers have analyzed "social presence", the feeling of being together with other users within the virtual space [4]. Several studies have shown how increased social presence in VR generated emotional reactions very similar to those of real interactions [20].

The use of virtual reality for training activities was already being studied in 1999 [18]. Factors that contribute to situations with elevated risks involve inadequate safety knowledge among on-site employees and a deficiency in safety consciousness and instruction. Broadly speaking, it's widely acknowledged that human mistakes are the primary factor behind numerous injuries within the engineering field.

Risky activities are generally conducted by following well-defined procedures but learning them in the real environment may be dangerous. The use of meta-verse for safety training lets the workforce learn about procedures. As an example, VR can be an appropriate tool for training roofing workers due to realistic perception of high-altitude situations without falling risks [8]. It also offers a high level of flexibility for constructing experiences that have been difficult to generate using other techniques, such as collaborating for conducting a submarine.

3 Collaborative Virtual Space

In this section, we describe the roles of the users and the collaborative virtual space implemented in MetaCUX.

3.1 The Roles

A user can play two roles in the system: (i) *organizer*, enabled to create a new public or private room, select the scenario, and manage the creation and scheduling of various activities, such as meetings, interviews, etc., (ii) *partici-pant*, enabled to enter rooms and perform activities organized by other users. In the Metaverse, users are represented by their avatars, and to ensure easy recognition among users, each avatar is flanked by a label, indicating the username. To facilitate collaboration among several users in the same environment we added an audio icon on the head of the avatar who is speaking.

3.2 Immersive Interaction Design and Implementation

The environment replicates a 3D collaborative environment, including 4 scenarios: training room, interview room, collaboration room, and simulation room.

Training Room. The "Training Room or Classroom" scenario (Fig. 1a) provides a dynamic and familiar simulation of a school classroom. It clearly defines roles, with students at desks and the teacher stationed near the whiteboard. The teacher can conduct lessons with real-time exercises on the board. The scenario features an interactive whiteboard (IWB) for presentations and videos.

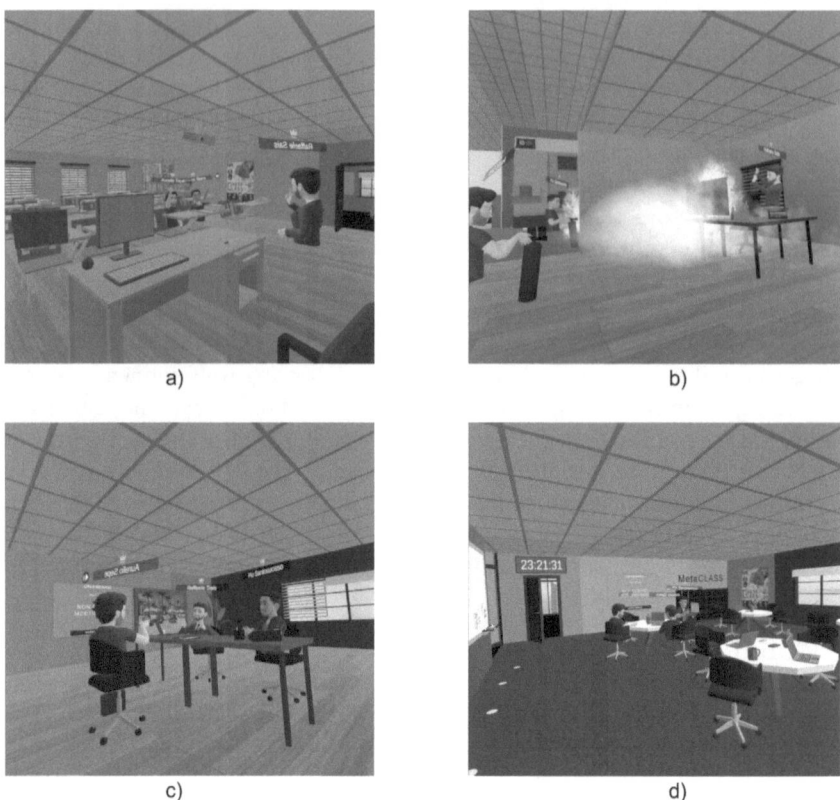

Fig. 1. The four scenarios: (a) Training Room, (b) Simulation Room, (c) Interview Room, and (d) Collaboration Room.

The furnishings, including books, a globe, and a clock, create a realistic learning environment, ensuring a strong connection between the virtual and real world.

Simulation Room. The "Simulation Room" scenario (Fig. 1b) is an extension of the training room, allowing for practical application of concepts learned. It facilitates a swift transition from theory to practice, saving valuable time. The scenario emulates a fire station, complete with a fire truck in a large garage. Various elements like a car, television, and office are simulated to be on fire, providing opportunities for practice. Functional fire extinguishers release gas to interact with the flames for realistic extinguishing. A panel allows for restarting the simulation. Additionally, an interactive whiteboard is available for video demonstrations on fire extinguisher use.

Interview Room. The "Interview Room" scenario (Fig. 1c) replicates a standard work environment interview between two individuals. It features a desk for direct interaction between the candidate and interviewer, with minimal distractions. The room's minimalist design ensures focus. Additionally, there are sofas for observers and a clock to keep track of time.

Collaboration Room. The "Collaboration Room" (Fig. 1d) represents a virtual space designed for collaborative meetings, fostering an informal and stimulating environment. It features a circular table layout to enhance interaction and communication. Each table is equipped with interactive tools like tablets and digital markers for group work. The design is inspired by leading tech companies like Google, Microsoft, and Apple, creating a relaxed atmosphere to boost creativity and productivity. The room also includes an interactive whiteboard for discussing drafts and ideas with team members.

3.3 The System Architecture

The system environment has been developed by using Unity[2], the most popular Game Engine for video game software development since the scope of virtual reality is very similar to video game development. The support of the multi-user features has been offered by the Unity-compatible Pun2 library, enabling multiple users to connect on the network to share the same virtual room. The Photon-Voice library creates an audio communication channel that allows a shared voice chat system among all users within the room. By taking advantage of the avatar SDK provided by Oculus, each user can reuse their own avatar, already created and customized within the Oculus account, within the platform. By synchronizing the Photon-Voice library with Avatars, it was possible to simulate the synchronization of the avatar's mouth movement while the user is speaking. To interact with the scenario and the various objects in it, the Oculus Interaction SDK was used to take full advantage of the technology and features offered by oculus visors, such as the ability to use HeadSet cameras for live hand tracking without using controllers. This feature, in addition, allows the user to record custom poses to simulate grasping objects within virtual reality. The XR Interaction Toolkit was used for interactions with interfaces and character movement, both via the left controller analog and via teleportation in order to decrease motion sickness. To simplify access, accounts already present on all Oculus devices have been leveraged, allowing the user not to have to register on a new platform, but to be able to take advantage of what is already available.

The architecture of the proposed system is depicted in Fig. 2. It is based on the Client-Server pattern: the ServerBackend developed using NodeJs manages data related to rooms, meetings, and permissions. It manages the application logic and the persistent data stored in a MySQL database. It answers the client's requests, where the client is an Oculus (shown in Fig. 3) equipped with the Android operating system and the MetaCUX app. The Client-ServerBackend communication is based on the HTTPS protocol. The data collected by the Oculus clients are managed by two cloud servers, shown in the leftmost part of Fig. 2: (i) Photon cloud, responsible for the management of the multiuser connections

---

[2] www.unity.com.

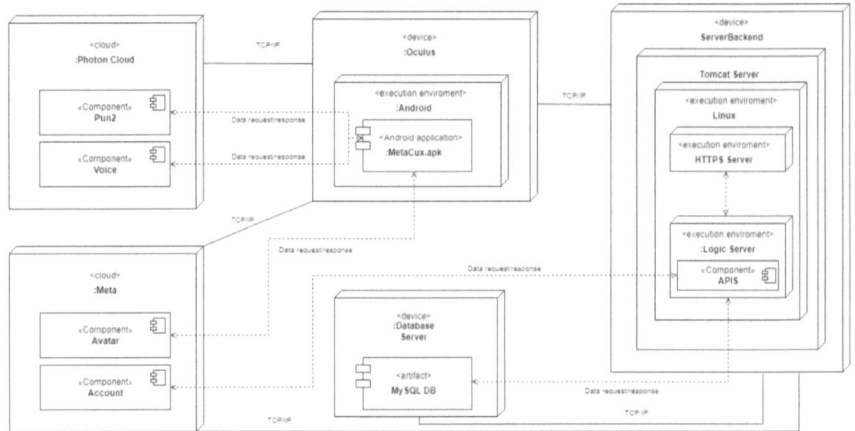

Fig. 2. The UML deployment diagram of the system.

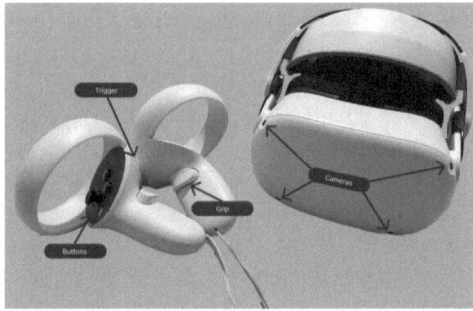

Fig. 3. The system client: the Oculus device.

among several clients of the same application. In particular, it performs the room management (Pun2 library) and vocal channels (Voice library). (ii) Meta. Useful to obtain information on the user's account, such as the user's avatars.

The ServerBackend has in charge the exchange of the oculus account information.

4   Design of the Training

In this section, we illustrate how to carry out training in virtual reality using the 4 scenarios presented. Both participants and a tutor should be present within the virtual environment, to encourage social interaction in the scenarios created. The participants are guided by the tutor, who plays the role of organizer of the activity within the virtual scenario: she observes the participants completing the tasks on the screen, helping them when they are in difficulty.

Pre-condition. Participants must practice with the devices and controllers to understand how to use buttons, interact with objects and interfaces, and move within the environment and between rooms. In the scenario presented for the simulation, a button on the controller is dedicated to removing the safety and operating the fire extinguisher jet. This pre-training phase ends when all participants felt comfortable enough with the Meta Quest 2 device.

Step 1: Training Room. In the training room, the organizer employs an interactive whiteboard to deliver a presentation on the proper utilization of fire extinguishers within the framework of company safety. The attendees attentively watch, seated at their respective desks.

Step 2: Simulation Room. Afterward, all users proceed to the simulation room, where, under the organizer's guidance, participants work together to extinguish the ignited objects.

Step 3: Collaboration Room. Following the simulation, all attendees transition to the Collaboration room, where they engage in a discussion about the hands-on experience they've just undergone. Each person has the opportunity to jot down the advantages and disadvantages of this activity on the chalkboard.

Step 4: Interview Room. At last, participants move into the interview room. One by one, each user is welcomed inside, engaging in a conversation with the organizer about his virtual reality experience and, in particular, the collaborative task simulation within the Metaverse.

Post-condition. At the end of the activity, once they leave the virtual environment, the participants should fill out a post-experiment questionnaire. This evaluation is intended to be qualitative as well as quantitative.

4.1   A Template of Evaluation

When designing a virtual experience simulating risky activities there is the need to assess whether participants are able to perform correctly their tasks with a user experience of good quality. Thus, we propose an evaluation template that may be useful to conduct user studies. Before starting the evaluation, we recommend collecting demographic data through a questionnaire and having participants sign the consent for using the data. An example of quantitative evaluation involves analyzing the number of errors, for example, a user pressing the wrong grip button on the analog controller or following the wrong paths, during observed activities. During the experiment, suitable time is given to carry out all activities. The tutor who monitors each participant may note the time taken to reach each room, the time taken to hold the fire extinguisher, and to hold the whiteboard pen.

After having conducted the experience we suggest the adoption of a standard questionnaire [21] for the qualitative evaluation. The questionnaire collects the participants' user experience by considering the following subscales: Presence, Involvement, Immersion, Usability, Ability, Flow, Emotion, and Consequence of the experience.

We also suggest evaluating social presence. To this end, we consider the five questions proposed in [13]. Example questions are: "I felt like I was in the presence of another person in the virtual environment." and "I felt that people in the virtual environment were aware of my presence."

As in [13], questions are rated on a seven-point Likert scale (from -3 = completely disagree to 3 = totally agree).

5   Conclusion

In this paper, we presented the design of virtual fire extinguisher training in the MetaCUX virtual environment, offering four main scenarios: training room, simulation room, interview room, and collaboration room. MetaCUX offers interaction and collaboration capabilities within the same environment. We also proposed a template for the experience assessment, consisting in the verification of the environment effectiveness and in the evaluation of the user experience.

This training design could interest researchers who want to further investigate the use of Metaverse and experiment with collaborative learning and working with the proposed interaction modes. Professionals can also use similar tools to carry out safety collaborative tasks.

In the future, we plan to conduct a case study with a large sample of participants to evaluate how the real and virtual training experience compares.

Acknowledgements. We acknowledge the course participants and the teacher. We also acknowledge financial support from the Research Projects of Significant National Interest (PRIN) 2022 PNRR, project n. D53D23017290001.

References

1. Ariyadewa, P., Wathsala, W., Pradeepan, V., Perera, R., Atukorale, D.: Virtual learning model for metaverses. In: 2010 International Conference on Advances in ICT for Emerging Regions (ICTer), pp. 81–85. IEEE (2010)
2. Barra, P., et al.: Demo: Metacux - a multi-user, multi-scenario environment for a cooperative workspace. In: Proceedings of the 15th Biannual Conference of the Italian SIGCHI Chapter. CHItaly 2023. Association for Computing Machinery, New York (2023). https://doi.org/10.1145/3605390.3610820. https://doi.org/10.1145/3605390.3610820
3. Barra, P., et al.: Metacux: Social interaction and collaboration in the metaverse. In: Abdelnour Nocera, J., Kristín Lárusdóttir, M., Petrie, H., Piccinno, A., Winckler, M. (eds.) INTERACT 2023. LNCS, pp. 528–532. Springer, Cham (2023). https://doi.org/10.1007/978-3-031-42293-5_67

4. Brondi, R., et al.: Evaluating the impact of highly immersive technologies and natural interaction on player engagement and flow experience in games. In: Chorianopoulos, K., Divitini, M., Hauge, J.B., Jaccheri, L., Malaka, R. (eds.) ICEC 2015. LNCS, vol. 9353, pp. 169–181. Springer, Cham (2015). https://doi.org/10.1007/978-3-319-24589-8_13

5. Cantone, A.A., et al.: Contextualized experiential language learning in the metaverse. In: Proceedings of the 15th Biannual Conference of the Italian SIGCHI Chapter, pp. 1–7 (2023)

6. De Lucia, A., Francese, R., Passero, I., Tortora, G.: Slmeeting: supporting collaborative work in second life. In: Proceedings of the Working Conference on Advanced Visual Interfaces, pp. 301–304 (2008)

7. Duan, H., Li, J., Fan, S., Lin, Z., Wu, X., Cai, W.: Metaverse for social good: a university campus prototype. In: Proceedings of the 29th ACM International Conference on Multimedia, pp. 153–161 (2021)

8. Ergun, H.: Monitoring physiological reactions of construction workers in virtual environment: a feasibility study using affective sensing technology (2015)

9. Evans, B.: The zoom revolution: 10 eye-popping stats from tech's new superstar. Cloud Wars (2020)

10. Fracaro, S.G., et al.: Towards design guidelines for virtual reality training for the chemical industry. Educ. Chem. Eng. **36**, 12–23 (2021)

11. García-Pereira, I., Vera, L., Aixendri, M.P., Portalés, C., Casas, S.: Multisensory experiences in virtual reality and augmented reality interaction paradigms. In: Smart Systems Design, Applications, and Challenges, pp. 276–298. IGI Global (2020)

12. Lege, R., Bonner, E.: Virtual reality in education: The promise, progress, and challenge. Japan Association for Language Teaching Computer Assisted Language Learning Journal (JALT CALL Journal) **16**(3) (2020)

13. Makransky, G., Lilleholt, L., Aaby, A.: Development and validation of the multimodal presence scale for virtual reality environments: a confirmatory factor analysis and item response theory approach. Comput. Hum. Behav. **72**, 276–285 (2017)

14. Matcha, W., Rambli, D.R.A.: Exploratory study on collaborative interaction through the use of augmented reality in science learning. Procedia Comput. Sci. **25**, 144–153 (2013)

15. Motomatsu, H.: Virtual reality in the medical field. UC Merced Undergraduate Res. J. **7**(1) (2014)

16. Peters, P.: Google's meet teleconferencing service now adding about 3 million users per day. The Verge (2020)

17. Qiu, Y.H., Luo, X.J., et al.: Application of computer virtual reality technology in modern sports. In: 2013 Third International Conference on Intelligent System Design and Engineering Applications, pp. 362–364. IEEE (2013)

18. Rickel, J., Johnson, W.L.: Virtual humans for team training in virtual reality. In: Proceedings of the Ninth International Conference on Artificial Intelligence in Education, vol. 578, p. 585. Citeseer (1999)

19. Rokooei, S., Shojaei, A., Alvanchi, A., Azad, R., Didehvar, N.: Virtual reality application for construction safety training. Saf. Sci. **157**, 105925 (2023). https://doi.org/10.1016/j.ssci.2022.105925. https://www.sciencedirect.com/science/article/pii/S0925753522002648

20. Schuemie, M., Straaten, P., Krijn, M., Mast, C.: Research on presence in virtual reality: A survey. Cyberpsychology & behavior: the impact of the Internet, multimedia and virtual reality on behavior and society **4**, 183–201 (2001). https://doi.org/10.1089/109493101300117884

21. Tcha-Tokey, K., Christmann, O., Loup-Escande, E., Richir, S.: Proposition and validation of a questionnaire to measure the user experience in immersive virtual environments. Int. J. Virtual Reality **16**(1), 33–48 (2016)
22. Thorp-Lancaster, D.: Microsoft teams hits 75 million daily active users, up from 44 million in march. Windows Central (2020)
23. Xiong, J., Hsiang, E.L., He, Z., Zhan, T., Wu, S.T.: Augmented reality and virtual reality displays: emerging technologies and future perspectives. Light: Sci. Appl. **10**(1), 216 (2021)

Data-Efficient Student Profiling in Online Courses

Gianni Fenu, Roberta Galici, and Mirko Marras$^{(\boxtimes)}$

Department of Mathematics and Computer Science, University of Cagliari, Cagliari, Italy
{fenu,roberta.galici,mirko.marras}@unica.it

Abstract. Online courses in higher education have gained popularity, but students struggle with self-regulation in online learning. The absence of traditional classroom guidance due to limited educator oversight highlights the need for effective student profiling. Existing profiling methods focus on non-university contexts with data-rich platforms, leaving platforms like Moodle at a disadvantage. In this paper, we explore the creation of useful student profiles with limited data, often found in Moodle and similar platforms. We propose to adopt a clustering method based on eight key self-regulation variables: revision, progress, consistency, dedication, regularity, focus, and practicality. Across diverse online university courses, our experiments show that our approach effectively identifies meaningful profiles, even with limited data. These profiles also reveal unique demographics, providing insights into online learning behavior.

Keywords: Student Profiling · Online Course · E-Learning · Learning Environment · Clustering · Demographic Analysis · Moodle

1 Introduction

Online courses, especially during the pandemic, have become a prominent mode of learning, offering students the flexibility to complete coursework virtually and at their convenience. However, self-regulation remains a challenge, as students may struggle with effective engagement due to the absence of face-to-face guidance [5,10]. Understanding student behavior is therefore imperative.

Clustering techniques have found application in the analysis of student behavior across digital learning environments, encompassing blended courses, intelligent tutoring systems, educational games, and massive open online courses (MOOCs). For instance, within the context of blended courses, [4] explored patterns of effort regulation among university students, while [17] observed diverse consistency patterns over time. Research by [1] focused on changes in learning strategies within blended learning. In [16], the authors investigated planning, engagement, evaluation, and reflection using log data, whereas [2] delved into the impact of student regularity on academic performance in MOOCs. Within the domain of MOOCs, [9] examined student commitment and consistency, and [6] identified student groups based on help-seeking behavior. Nevertheless, none of these prior studies has addressed the challenge of clustering multiple learning

F. Palomba and C. Gravino (Eds.): WAILS 2024, LNCS 14545, pp. 11–20, 2024.
https://doi.org/10.1007/978-3-031-57402-3_2

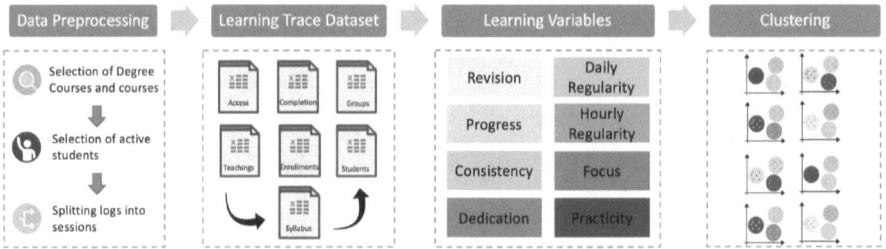

Fig. 1. Our methodology involves three main stages: data preprocessing, conversion into tabular data, learning variables extraction, and clustering.

variables in the context of online university courses, particularly where data collection is constrained. Widely adopted platforms, such as Moodle, often lack the capacity to capture detailed student behavior data, such as video play or pause events, thus complicating the creation of meaningful student profiles.

In this paper, we explore the adoption of methods to extract student profiles within the context of online university courses, under scenarios that suffer from technical limitations in data sources, as observed in Moodle-based platforms. To accomplish this goal, we analyze trace data acquired from student interactions with an online learning platform based on Moodle, spanning across a diverse range of online courses. Our approach entails the systematic collection and preprocessing of trace data, followed by the application of clustering techniques to multiple learning variables. Subsequently, we create and inspect the clusters, highlighting their distinct learning patterns and demographic compositions. With this study, our primary aim is to address two research questions: (**RQ1**) Can educationally meaningful student profiles be derived effectively in the context of online university courses, under limited data collection? (**RQ2**) To what extent do student profiles exhibit tendencies that correlate with specific demographic attributes, such as gender, age, and geographical origin?

Our comparative analysis of online university courses throughout a semester shows that it is indeed possible to identify and interpret student profiles based on cluster centroids, even in this challenging context. Simultaneously, our analysis reveals noteworthy disparities in demographic compositions across these profiles.

2   Methodology

In this section, we provide an overview of the methodology employed in our study. First, we present the learning environment and describe the dataset obtained from the Moodle platform adopted by the considered University. Subsequently, we introduce the clustering pipeline implemented for our analysis, as depicted in Fig. 1. Each step of the pipeline is detailed in the following subsections.

2.1   Learning Environment

Our study analyzed trace data from the University's Moodle platform, commonly used for online courses. Students accessed resources within the courses they were enrolled, which mainly included video lectures and practical exercises. The video lectures were provided as SCORM packages[1], facilitating web-based learning content interoperability, accessibility, and reusability. Specifically, these packages included videos, slides, and a bibliography. Practical exercises included hands-on tasks with instructor-provided guidelines. Typically, students downloaded and completed the exercises, finally uploading their solutions to Moodle.

Our study focused on the courses delivered online by the considered University during the initial phase of the pandemic (March-July 2020). These courses, which had already been delivered online prior to the pandemic, offered video lectures and practical exercises. For conciseness, we specifically concentrated on four Bachelor's degree programs: *Administration* (13 courses), *Cultural Heritage* (10 courses), *Economics* (14 courses), and *Engineering* (16 courses). Their courses utilized Moodle for teaching, while other courses, although online, followed a synchronous learning approach using Adobe Connect [7]. It is important to note that each degree program's course was attended by students in their first, second, and third years, as these programs span three years.

2.2   Learning Trace Dataset

The learning trace data prioritizes student privacy and data anonymization, adhering to privacy regulations. It records interactions by anonymized students on the platform. These interactions include actions like accessing the course main page, opening SCORM packages, and submitting exercises. For example, students start by viewing the course page (event: "Course Displayed"), then engage with course resources like SCORM packages (event: "Course Module Displayed"). They may launch these packages (event: "SCO launched") and complete activities, recorded as "Sent Status SCO". Students also perform actions like submitting assignments (event: "Submission module view") and checking homework status (event: "Homework, Displayed delivery status"). Notably, due to local privacy constraints, students' final exam grades were inaccessible. The trace data collected from Moodle has a relatively coarse granularity, posing challenges in accurately identifying student profiles based on course behavior.

The dataset originally consisted of six key tables. The "Students" table contains student demographic information, including anonymized user identifiers, their degree programs, gender, age, and region of origin. The "Courses" table provides details about courses, such as course identifiers, program affiliations, and year in the degree program. The "Schedule" table records course resource information, including resource identifiers, associated courses, resource titles, and types (lecture or exercise). The "Enrollments" table lists enrolled courses with unique user and course identifiers. The "Access" table monitors student interactions with the Moodle platform, logging timestamps, user identifiers, involved course and resource, actions taken (e.g., "Course Displayed"), textual descriptions (e.g., "User with ID 12 viewed course with ID 24"), and the device used

---

[1] SCORM stands for Sharable Content Object Reference Model, and it is a set of standards and specifications for creating and packaging e-learning content. A SCORM package typically includes a collection of web-based learning resources (such as presentations, quizzes, and simulations) that adhere to the SCORM standards.

Table 1. Courses in our study, sorted by year and number of interactions.

Year	ID (Name)	#Interactions	#Students	#Packages	#Exercises
1	ALG (Algebra)	118,559	918	24	–
	PHY (Physics)	126,770	931	40	27
	EC (Electronic Computers)	66,181	458	48	–
	MC (Mathematics)	17,438	198	27	–
	ECS (Elements of Computer Science)	36,989	426	24	–
2	EE (Electrical Engineering)	56,772	593	69	7
	SA (Systems Analysis)	37,234	223	27	5

(e.g., "web" or "mobile"). Lastly, the "Completion" table indicates if a student completed a given course activity, also including timestamps, user identifiers, and course and resource identifiers. These tables represented the core of our dataset, enabling the analysis of student behavior in the learning environment.

2.3 Data Preprocessing

In our pipeline, we performed three main preprocessing steps. Initially, we focused on identifying degree programs and associated courses that heavily relied on Moodle as their primary teaching platform during the specified time-frame. To accomplish this, we ranked courses within each degree program based on the average number of interactions per student. We exclusively considered courses with an average of more than 100 interactions per student and degree programs with a minimum of 4 courses meeting this criterion. This selection process aimed to ensure that we had a substantial volume of interactions available to construct meaningful student profiles. Second, to address the challenge posed by the coarse granularity of trace data, we removed students with a total number of interactions lower than the average number of interactions per student within their respective courses. This step allowed us to retain students who exhibited a sufficient level of engagement to enable meaningful profile construction. In the final pre-processing step, we organized student logs into sessions based on a defined criterion. Specifically, we considered two consecutive interactions as belonging to distinct sessions if the time interval between them exceeded 30 min. This threshold, widely employed in prior studies, e.g., [12], accounts for cases where logout events are not explicitly recorded, as ours. Table 1 provides an overview of courses characterized by the highest average interactions per student, all belonging to the same degree program (*Engineering*). We will focus our analysis on ECS and SA courses due to their diverse characteristics, including study duration, topics, and course structures. In Fig. 2, we depict the demographics of students in these courses based on their gender, birth year, and geographic origin. Percentages represent the entire student population before pre-processing.

2.4 Learning Variables

The concept of self-regulated learning (SRL) in digital learning environments has been a subject of extensive study [3, 17]. These investigations have revealed that learners can be characterized based on their patterns of online engagement.

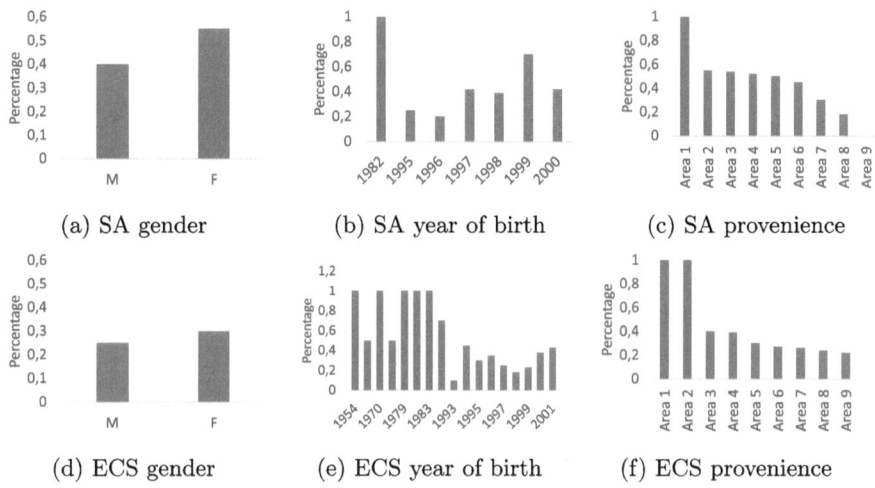

(a) SA gender   (b) SA year of birth   (c) SA provenience

(d) ECS gender   (e) ECS year of birth   (f) ECS provenience

Fig. 2. Gender, provenience, and birth year in SA (top) and ECS (bottom).

Based on these insights, we identified eight learning variables to gain deeper insights into student SRL behavior within the considered learning environment.

Revision assesses a student's propensity to review course content, particularly in preparation for final exams, by calculating the percentage of lessons viewed in a given week that were previously accessed. Progress is crucial for grasping student's engagement [8], with this variable quantifying the percentage of completed activities out of the total course activities (i.e., a proxy of their overall course involvement). Consistency measures the stability of a student's time allocation over different weeks, distinguishing consistent students from those who fluctuate. Dedication reflects student's commitment, determined by the number of course activities completed in a specific week. Daily Regularity observes systematic access patterns on specific days of the week, such as Mondays, Tuesdays, and Fridays, calculating the percentage of students adhering to this routine. Hourly Regularity monitors consistent access times in the day, often due to exercise submission deadlines. Focus evaluates how students allocate their study time among various activities within a week, assessing even distribution or concentration on specific tasks. Lastly, Practicity quantifies the percentage of weekly activities related to exercises (therefore excluding lecture-based tasks), further enhancing the understanding of students' SRL behavior in the online learning environment. These eight learning variables form a comprehensive framework for analyzing students' SRL in online courses.

2.5   Clustering

To create the student profiles, we employed a clustering approach based on the K-Means algorithm [11], identifying patterns in student behavior. Each learning variable was independently subjected to clustering, treating students as vectors with a length equal to the number of course weeks. These vectors capture the specific variable values for each student in each week. The clustering process was essential for grouping students with similar behavioral patterns (i.e., distinct profiles of student). To determine the optimal number of clusters (K), we leverage the Silhouette score [15], where higher scores signify better cluster separability. We explore K values from 2 to 10 and select the value that maximizes the Silhouette score (0.72 at K=3). Identified the optimal number of clusters, our process assigns a cluster label to each student. The resulting clusters represent cohorts of students with analogous learning behaviors. Being based on learning indicators, our approach not only enhances profiles interpretability but also contributes to their robustness.

3   Experimental Results

For each selected course, we first analyzed and compared the obtained profiles (RQ1) and then investigated their demographics composition (RQ2).

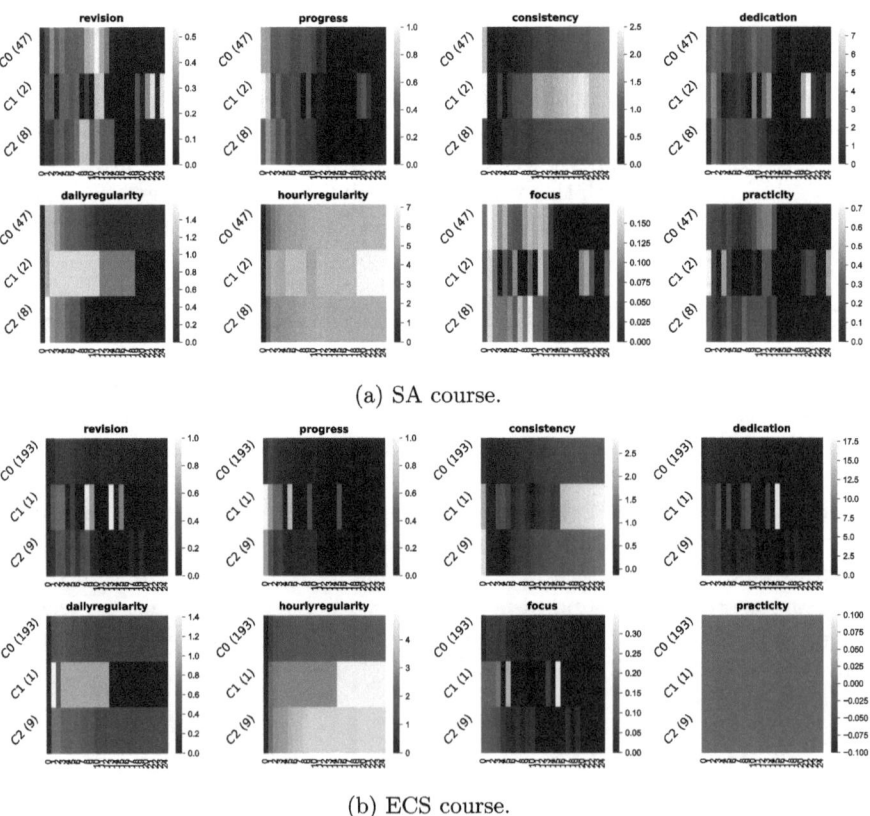

(a) SA course.

(b) ECS course.

Fig. 3. Clusters' centroids throughout course weeks.

3.1   RQ1 Learning Profiles

In a first analysis, we investigated whether it is possible to extract educationally meaningful student profiles in online university courses under data collection with limited source of evidence. Though we clustered students in each selected course, we report and analyze the most significant ones in detail (ECS and SA).

Figure 3b reports the cluster centroids patterns for each learning variable in the ECS course. Specifically, three clusters can be observed: C0 with 193 students, C1 with 1 student, and C2 with 9 students. The plots show 24 weeks, however, the lectures in each course lasted 12 weeks. We decided to show a larger number of weeks to investigate whether any behavioral difference emerges during the exam preparation period. C0 and C2 have a similar behaviour in progress, dedication and focus from the second week of lesson to the end of the course. On the contrary, the consistency is very low, in particular for the first weeks and it increases at the end, especially for C0. Daily regularity is very high for the cluster C0 and very low for C1 and C2. Hourly regularity have a high percentage in C0 from the 4th week. However, in C2, we have a medium percentage and it is similar to C1. Revision behaves similarly across all three clusters. In fact, we have a percentage equal to zero for the first week. Afterwards we have a high percentage and then again a very low one. Finally, the practicity assumes percentage values equal to 0 for each cluster and for each week. This probably means that there were too few exercises for this specific course.

In Figure 3a we have the course of SA, a 2nd year course, with three distinct clusters: C0 with 20 students, C1 with 46 students, and C2 with 32 students. In particular, the first cluster (C0) has a similar behaviour to the last cluster (C2). Revision, focus and dedication have similar patterns and the percentages are higher at the beginning of the course. Daily regularity is higher at the beginning and then it becomes very low. On the contrary, hour regularity is low at the beginning and high at the end; therefore it increases over time, for C0, C1, and C2. Progress is low in all weeks, except for the first week, which is very high for all clusters. Consistency has low values at the beginning; then, it increases over the weeks for all three clusters. The only difference is observed at the week 0, on which we reported a percentage of 0 for C0, a percentage of 0.2 for C1, and 0.6 for C2. Finally, the practicity seems to be very high only in some specific week, such as week 10, 11, 12 and 13 for the cluster C2. This might be due to students trying to do more exercises when they are closer to the final exam.

Findings RQ1. *In summary, our analysis demonstrates the potential to extract valuable student profiles in online university courses, even when data sources are limited. These profiles can serve as a source of reasoning for educators and as targets for customized interventions.*

3.2   RQ2 Demographic Analysis

In a second analysis, we investigated whether there is a tendency of certain profiles than others to be associated with specific demographic classes, considering gender, age, and geographic provenience as the main factors. To this end, we represent the distribution of demographic classes per cluster in Figure 4. In particular, we present, in order, the clusters of the gender, the age groups and the provenience of the students that participates in that particular course (SA and ECS). The analysis clearly reveals a predominant female representation in the SA course, as depicted in Figure 4a, whereas the ECS course exhibits a notable male majority, illustrated in Figure 4d. However, there are no substantial differences in gender representation across clusters, except for C0 in ECS (having lower consistency and hourly regularity). As for the age representation, it seems that in general the courses included students who are on track within the degree program, with most of the students who attended SA (ECS) having 1999 (2000) as the year of birth (Fig. 4b and 4e). Interestingly, C0 in SA included a larger part of those on-track students, which were associated with higher revision, dedication, and focus, among others, at the very beginning of the course. Observations can be made on ECS for C1, which included students who were consistent and regular towards the exam preparation period; nevertheless, it should be noted that, in this case, the cluster is small. Finally, we found interesting patterns also on the geographic provenience of students across clusters; it is worth noticing that this factor might be a proxy of the background. In the SA course (Fig. 4c), there are certain areas, such as Area 4 and 6, where students in C2 (with increasing overall engagement in the course) are predominant. Geographic provenience is a key factor also for clusters in ECS (Fig. 4f).

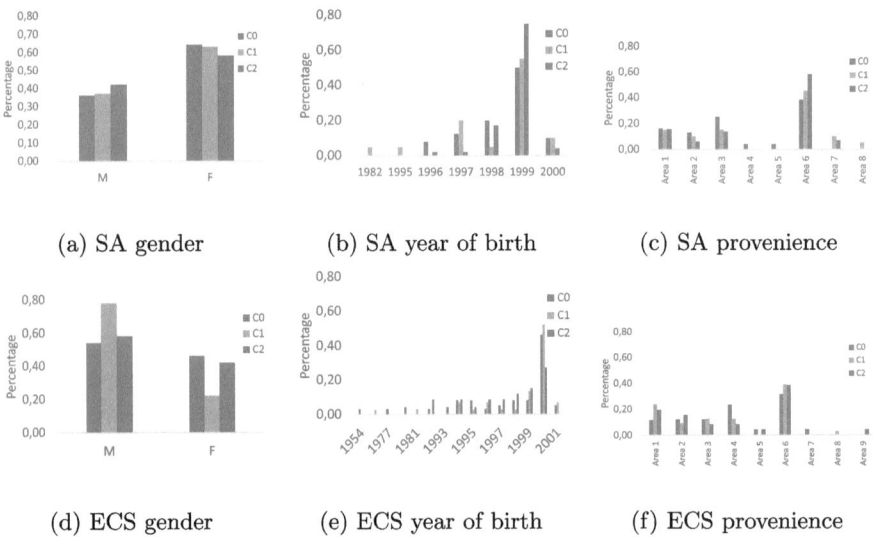

(a) SA gender   (b) SA year of birth   (c) SA provenience

(d) ECS gender   (e) ECS year of birth   (f) ECS provenience

Fig. 4. Student representation in clusters in SA (top) and ECS (bottom) courses.

Findings RQ2. *To sum up, our analysis revealed that certain profiles tend to be characterized by specific demographic classes. Such insight can be leveraged by educators to increase their knowledge about the context and adjust their teaching delivery according to this demographic pattern.*

4 Discussion and Conclusions

This study involved the analysis of various learning variables to derive interpretable student profiles with limited data (**RQ1**). We also explored the relationship between these profiles and demographic factors (**RQ2**). Unlike prior work focused on single courses [13,14], our approach considered multiple courses.

Our results show that our pipeline often resulted in three distinct clusters (**RQ1**). In general, the majority of students exhibit increased activity levels at both the beginning and the end of the course, particularly as the exam session approaches. Importantly, our findings demonstrate the successful application of student profiling within learning environments, such as Moodle, even in cases where fine-grained data collection is often unfeasible (unlike [12]). These findings highlight the potential of student profiling in privacy-constrained scenarios, where detailed tracking with video content may not be feasible. Our analysis revealed that certain profiles tend to be characterized by specific demographic classes, especially based on the age and geographic provenience (**RQ2**). In particular, there are clusters, associated with particular demographics, that seem to be related to students who regularly follow lectures. For both courses analyzed, we noticed that there are imbalances with respect to the gender, that already shows up in the original enrolments from the selected degree program.

In conclusion, our method aims to enhance student profiling in scenarios with limited behavioral evidence, facilitating personalized interventions in education. However, our study has certain limitations. The restricted availability of courses on the Moodle platform may impact the generalizability of findings. Future research will expand the scope to include a more extensive and diverse sample of courses. Privacy constraints and the unavailability of detailed data, such as video playback events and final exam grades, limit the depth of our analyses. Similarly, students' background could be an interesting variable that could be correlated to the tendencies of each student. In general, considering a control variable will help to assure the students are not affected by this confounding variable. a possible concern is related to the selection of the data to build the K-means model. Limitations addressed in future work pertain also to the amount of data, which could not be enough to lead to an accurate modelling. Similarly, the adopted splitting might not allow to consider that students tendency could be correlated to the course itself. Ongoing efforts will finally involve a more comprehensive demographic analysis and a larger period of interest.

References

1. Barthakur, A., Kovanovic, V., Joksimovic, S., Siemens, G., Richey, M.C., Dawson, S.: Assessing program-level learning strategies in Moocs. Comput. Hum. Behav. **117**, 106674 (2021)
2. Boroujeni, M.S., Sharma, K., Kidziński, Ł, Lucignano, L., Dillenbourg, P.: How to quantify student's regularity? In: Verbert, K., Sharples, M., Klobučar, T. (eds.) Adaptive and Adaptable Learning, pp. 277–291. Springer International Publishing, Cham (2016). https://doi.org/10.1007/978-3-319-45153-4_21
3. Cho, M.-H., Cheon, J., Lim, S.: Preservice teachers' motivation profiles, self-regulation, and affective outcomes in online learning. Dist. Ed. **42**(1), 37–54 (2021)
4. Cho, M.-H., Shen, D.: Self-regulation in online learning. Distance Educ. **34**(3), 290–301 (2013)
5. Chunkhare, M., Jadhav, S.: "online learning" technology solutions during the COVID-19 pandemic: An empirical study of medical technology and allied health-care student perceptions. Int. J. Virtual Pers. Learn. Environ. **13**(1), 1–11 (2023)
6. Corrin, L., de Barba, P.G., Bakharia, A.: Using learning analytics to explore help-seeking learner profiles in MOOCs. In Proc. of LAK **2017**, 424–428 (2017)
7. Fenu, G., Galici, R.: Modelling student behavior in synchronous online learning during the Covid-19 pandemic (2021)
8. Hew, K.F., Qiao, C., Tang, Y.: Understanding student engagement in large-scale open online courses: a machine learning facilitated analysis of student's reflections in 18 highly rated MOOCs. Int. Rev. Res. Open Distrib. Learn. **19**(3) (2018)
9. Khalil, M., Ebner, M.: Clustering patterns of engagement in massive open online courses (MOOCs): the use of learning analytics to reveal student categories. J. Comput. High. Educ. **29**(1), 114–132 (2017)
10. Kotsiantis, S.B., Pierrakeas, C.J., Pintelas, P.E.: Preventing student dropout in distance learning using machine learning techniques. In: Palade, V., Howlett, R.J., Jain, L. (eds.) Knowledge-Based Intelligent Information and Engineering Systems, pp. 267–274. Springer Berlin Heidelberg, Berlin, Heidelberg (2003). https://doi.org/10.1007/978-3-540-45226-3_37
11. Likas, A., Vlassis, N., Verbeek, J.J.: The global k-means clustering algorithm. Pattern Recogn. **36**(2), 451–461 (2003)
12. M. Marras, J. T. T. Vignoud, and T. Käser. Can feature predictive power generalize? benchmarking early predictors of student success across flipped and online courses. In: Proceedings of EDM 2021 (2021)
13. Matcha, W., Gašević, D., Uzir, N.A., Jovanović, J., Pardo, A.: Analytics of learning strategies: Associations with academic performance and feedback. In Proc. of LAK **2019**, 461–470 (2019)
14. Pardo, A., Gašević, D., Jovanovic, J., Dawson, S., Mirriahi, N.: Exploring student interactions with preparation activities in a flipped classroom experience. IEEE Trans. Learn. Technol. **12**(3), 333–346 (2018)
15. Rousseeuw, P.J.: Silhouettes: a graphical aid to the interpretation and validation of cluster analysis. J. of Comp. and App. Math. **20**, 53–65 (1987)
16. Saint, J., Whitelock-Wainwright, A., Gasevic, D., Pardo, A.: Trace-srl: a framework for analysis of microlevel processes of self-regulated learning from trace data. IEEE Trans. Learn. Technol. **13**(4), 861–877 (2020)
17. Sher, V., Hatala, M., Gasevic, D.: Analyzing the consistency in within-activity learning patterns in blended learning. In: Proceedings of LAK, pp. 1–10 (2020)

Exploring Artificial Intelligence in Videogames: An Application to the Didactic Paradigm "the Virtual Class"

Veronica Beatini[✉], Lucia Campitiello, Stefano Di Tore, and Maurizio Sibilio

Department of Humanities, Philosophy and Education, University of Salerno,
Fisciano, Italy
veronicabeatini@gmail.com

Abstract. Artificial intelligence (AI) has transformed the world of videogames, making them more immersive and engaging experiences.

Many studies explore the multiple ways AI has revolutionized the world of videogames. AI's ability to create more realistic and responsive non-player characters (NPCs) has made virtual worlds more immersive.

In addition to creating engaging game worlds, AI has contributed to the procedural design of levels and scenarios, reducing the workload of developers and ensuring greater variability. In this study, we aim to demonstrate how the use of AI is beneficial for creating new combinations of scenarios within our videogame "the virtual class", a didactic paradigm to measure perspective taking ability. Using AI within our videogame allows to change environments and scenarios in a way that enhances spatial dynamism allowing children to change the strategies used to answer. This could lead to a new version of the videogame that emphasizes customization and enhances children's engagement during the game.

Keywords: Artificial Intelligence · Videogame · Didactic

1 Introduction

1.1 AI and Videogames

Artificial Intelligence, also known as AI, is a rapidly growing field that aims to create machines and software that can simulate human intelligence at the intersection of computer science, mathematics, and engineering [1]. AI systems are designed to perform tasks that typically require human intelligence, such as learning from data, making decisions, solving problems, and understanding natural language. Over the years, AI has made remarkable strides, enabling machines to perform a wide range of functions, from speech recognition and image analysis to autonomous driving and medical diagnosis [2]. The development of AI continues to have a profound impact on various aspects of our lives,

F. Palomba and C. Gravino (Eds.): WAILS 2024, LNCS 14545, pp. 21–26, 2024.
https://doi.org/10.1007/978-3-031-57402-3_3

from enhancing productivity and automating repetitive tasks to revolutionizing industries and sparking discussions about the ethical and societal implications of intelligent machines [2]. For example, the videogame industry has undergone a profound transformation thanks to the increasing use of AI making them more immersive and engaging experiences [3,4]. Developers employ AI to bring non-playable characters (NPCs) to life, making them act intelligently and respond to the player's actions, thereby creating ever more sophisticated challenges [1]. Additionally, AI is essential for enhancing the gaming experience by real-time difficulty adjustment, game play personalization, and graphics enhancement [5].

1.2   Reinforcement Learning for Videogames

A promising machine learning method that is growing in popularity in the videogame industry in recent years is reinforcement learning. There are a variety of reinforcement methods to choose from knowing which method is suitable for which type of game is not easy. Training also takes time and it can be difficult to tell whether a training session was successful before closed it. In general, reinforcement learning is a paradigm of machine learning where an agent learns to take actions in an environment to maximize cumulative reward. Essentially, the agent makes decisions through continuous interaction with the environment, receiving feedback in the form of rewards or penalties based on its actions [6]. In the realm of gaming, reinforcement learning has been widely employed to train virtual agents in complex games. Reinforcement learning algorithms can learn optimal strategies to overcome specific challenges, dynamically adapting to the game conditions. For instance, an agent might learn to play a strategic videogame by adjusting its strategy based on opponents' moves and in-game conditions. This approach has also been utilized for training agents in real-time strategy games, role-playing games, and sports games, showcasing the versatility of reinforcement learning in addressing a wide range of gaming contexts. Reinforcement Learning tools such as The Unity machine learning toolkit is now even better suited for game developers [7]. Machine learning Toolkit makes it possible to train gaming AI in a powerful and modern way using reinforcement learning algorithms and has enabled the creation of more realistic and complex virtual worlds [7].

2   Tools

2.1   The Virtual Class: A Didactic Paradigm

The virtual class is a didactic paradigm developed by Di Tore (2020), using Unity3D framework with the purpose of assessing perspective taking (PT) in children aged 6 to 12 (Fig. 1). The game involves completing a task composed of two 3D scenes, aimed at measuring a child's ability to inhibit egocentric spatial coordinates and adopt another spatial perspective [8].

Specifically, a screen divided in two part is shown to the child. The first part frames the class from an overhead perspective, showing all the students, the teacher and the objects that make up the class; the second shows the perspective of one student. The child is asked to answer to which student belongs the point of view form which you see the image below. The game consists of two phases (approximate duration: 30 min): (a) a practice phase, and (b) a test phase. The test phase consists of 6 trials, and each lasts 15 s. Each experiment was conducted with the accompanying adult (parent, teacher, school assistant, or caregiver, as appropriate) and with a neuroscientist, psychiatrist, or psychologist from the department.

Fig. 1. Images from the videogame Virtual Class during the task

2.2   Unity ML-Agents Toolkit

Unity Machine Learning Toolkit (ML-Agents) is a powerful framework developed by Unity Technologies that enables the integration of machine learning algorithms with Unity, a popular game development engine. This toolkit is designed to facilitate the training of intelligent agents within Unity environments, allowing developers to create games with more realistic and adaptive behaviors.

The ML Agents toolkit consists of three parts main components: the agent, the environment, and the operations available for the agent. Agents share their experiences during training. Your strategy requires access to player input, scripts, and neural data networking via Python [9]. Commercial videogames offer significant advantages as learning environments because they feature agents that can be harnessed for educational purposes. These agents include characters and non-playable characters (NPCs). The games provide a rich environment in which agents can pursue micro or macro goals, such as achieving high scores or completing specific tasks. Agents can be configured to optimize their task completion and maximize rewards, all within a fully or partially observable state [10].

The observable state encompasses character statistics, such as health, attack power, and other attributes, along with NPCs whose health and characteristics also play a role in the environment. The main impediment to leveraging commercial videogames as learning environments lies in the fact that they are not open source. Consequently, researchers are unable to modify the source code to implement the necessary learning algorithms [11].

ML-Agents supports various machine learning techniques, including reinforcement learning, imitation learning, and evolutionary strategies. Developers can use this toolkit to train agents to navigate virtual environments, solve complex problems, or interact with game elements in a more sophisticated manner.

One of the key features of ML-Agents is its flexibility in defining custom reward structures, observations, and actions for the agents. This flexibility allows developers to tailor the learning process to specific game scenarios, making it applicable to a wide range of gaming applications.

By bridging the gap between machine learning and game development, Unity ML-Agents empowers developers to create more intelligent and dynamic virtual characters, enhancing the overall gaming experience. The toolkit has gained popularity for its accessibility and ease of integration, making it a valuable resource for those looking to incorporate machine learning capabilities into their Unity-based projects.

3   Virtual Class and ML-Agents: Vicariant Approach to Improve the Spatial Dynamics of the Game

The term 'vicariant approach' is not widely known in the context of game design or artificial intelligence in games. However, it could be a formulation that refers to 'multiple' approaches to get a solution, or to solve a problem. In the context of game design it could refer to different strategies to improve the spatial dynamics of videogames. [12] In general, to enhance the spatial dynamics of a game, developers might explore various strategies, including optimizing the graphics engine to handle a vast game world, implementing intelligent navigation techniques for non-player characters, or using procedural generation algorithms to create more varied and interesting environments.

In this study, we aimed to highlight how the use of AI, specifically the Unity ML-Agents toolkit, within the virtual classroom video game, is useful to create a dynamic game space system that serves educational purposes. The objectives of applying AI to the Virtual Class are multiple:

a) Using AI to create different combinations of the environment and encourage spatial dynamism. This may be necessary to understand various response strategies used by children [11].
b) Creating variability in game tasks and potentially customizing versions (including adjusting the difficulty level based on the user's answers) [13].
c) Creating with AI a vicariant tool for PT [14].

Once the objectives have been defined, our intention was not to provide a detailed explanation of how these algorithms are implemented within the video game but simply to demonstrate their effectiveness within it. Considering the close connection between the first and second goals, we believe that it is important to strive for recreating various combinations of the environment such as rotating the class, desks and students in order to observe the different strategies employed by users. We aim to demonstrate how effectively the use of AI enhances the ability to focus on the variability in users' behavior during a task. In both the field of Education and Cognitive Neuroscience, this strategic use of AI could be very useful for identifying individual characteristics based on their responses and establishing a customized learning path. In relation to the third goal, the concept of vicariance becomes extraordinarily relevant within the context of social references, as it signifies the adaptive process initiated by human beings to enhance decision-making and problem-solving. [15]. Human perception and behavior are linked to a reference environment to which behavior and thoughts adapt faced with a complex society that learns through technology-oriented models. Customizing a video game through the implementation of dynamic spatial strategies represents an innovative approach to provide a unique and engaging gaming experience. This customization can involve the creation of dynamic game worlds, where space adapts in real-time to the player's actions. For instance, modifying environmental structures or altering weather conditions based on player decisions contributes to maintaining interest and challenge. Additionally, the use of procedural space generation algorithms enables the creation of ever-changing environments, ensuring continuous variety and making each play-through unique. The integration of dynamic spatial strategies not only allows for greater game personalization but also adaptive responses to player preferences and skills, significantly enhancing the overall immersiveness of the virtual world created. To conclude, in our future research we will focus deeper on how human-machine interaction enhances the analysis of human behavior and the complex vicarious ways by which it expresses itself.

References

1. Russell, S.J., Norvig, P.: Artificial Intelligence: A Modern Approach. Pearson (2009)
2. Seidel, S., Berente, N., Lindberg, A., Lyytinen, K., Martinez, B., Nickerson, J.V.: Artificial intelligence and video game creation: a framework for the new logic of autonomous design. J. Digital Soc. Res. **2**(3), 126–157 (2020). https://doi.org/10.33621/jdsr.v2i3.46
3. Yannakakis, G.N.: Game AI revisited. In: Proceedings of the 9th conference on Computing Frontiers - CF '12, Cagliari, Italy, p. 285 (2012). https://doi.org/10.1145/2212908.2212954.
4. Yue, B., de Byl, P.: The state of the art in game AI standardisation. In: Proceedings of the 2006 International Conference on Game Research and Development, pp. 41–46, Murdoch University (2006)
5. Straeubig, M.: Games, AI and Systems. Eludamos J. Comput. Game Cult. **10**(1), Art. no. 1, Apr. 2020
6. Souchleris, K., Sidiropoulos, G.K., Papakostas, G.A.: Reinforcement learning in game industry-review, prospects and challenges. Appl. Sci. **13**, 2443 (2023). https://doi.org/10.3390/app13042443

7. Graham, J., Starzyk, J.A., Jachyra, D.: Opportunistic behavior in motivated learning agents. IEEE Trans. Neural Networks Learn. Syst. **26**(8), 1735–1746 (2015). https://doi.org/10.1109/TNNLS.2014.2354400
8. Viola, ilaria, Lecce, A., Campitiello, L., Di Tore, S.: La Perspective taking e i meccanismi di codifica in ambiente virtuale. J. Inclusive Methodol. Technol. Learn. Teach. **2**(1) (2022). Recuperato da https://inclusiveteaching.it/index.php/inclusiveteaching/article/view/8
9. Juliani, A., Berges, V. P., Teng, E., Cohen, A., Harper, J., Elion, C., Lange, D.: Unity: a general platform for intelligent agents (2018). arXiv preprint arXiv:1809.02627
10. Biçak, K.B.: An Evaluation of the Unity Machine Learning Agents Toolkit in Dense and Sparse Reward Video Game Environments (2021)
11. Kirsh, D.: The intelligent use of space. Artif. Intell. **73**(1–2), 31–68 (1995). https://doi.org/10.1016/0004-3702(94)00017-U
12. Berthoz, A.: La vicarianza. Codice, Torino (2013)
13. Mateas, M., Stern, A.: A behavior language for story-based believable agents. IEEE Intell. Syst. Their Appl. **17**(4), 39–47 (2002)
14. Sibilio, M.: La Didattica Semplessa. Liguori Editore, Napoli (2014)
15. Lake, B.M., Ullman, T.D., Tenenbaum, J.B., Gershman, S.J.: Building Machines That Learn and Think Like People, ArXiv160400289 Cs Stat, Nov. 2016, Accessed: Dec. 04, 2021

Reflections on the Implications of Artificial Intelligence in Inclusive Education

Amelia Lecce$^{(\boxtimes)}$

University of Sannio, Benevento, via dei Mulini, Italy
alecce@unisannio.it

Abstract. Artificial Intelligence is rapidly becoming a technological tool increasingly used in everyday life and presents itself as a means of improving people's social participation. However, its use raises countless issues related to ethics, transparency or data privacy. Leaving aside the interdisciplinary issues of the ethics of Artificial Intelligence, it is considered important to underline the potential of this technology in improving teaching action. Therefore, the purpose of this reflection is to consider the strengths of the use of Artificial Intelligence in the didactic practice of teaching and learning. In fact, these technological tools, if adopted in a constructive way, could become powerful allies in teaching interaction. For this reason, some teaching research is evaluating chatbot technology as a tool to improve the teaching experience. Chatbots, in fact, provide immediate answers to disciplinary questions and several studies suggest that adopting this technology in the classroom would offer students a more personalized, engaging and accessible learning environment.

Keywords: Inclusive Education · Innovative teaching · AI

1 The Opportunities of Artificial Intelligence in Inclusive Education

In the history of humanity, all technical-scientific revolutions have been accompanied by conflicting feelings that perpetuate prejudices and discrimination to the detriment of technologies. Last but not least, the current debate on the use of Artificial Intelligence (AI) tends to reproduce two sides: those who praise the infinite opportunities and those who instead warn about the risks.

AI tools and techniques are widely deployed across several industries and fields of science, however, the incalculable opportunities present significant challenges and ethical issues that must be addressed responsibly. Some of these challenges include transparency and interpretability of algorithms, data privacy, fairness and ethics in the use of AI. The UNESCO Recommendations on the use of AI are an invitation to political action by Member States, companies active in the education sector and teachers which in the context of "Education and research" translates as:

- in promoting awareness programs on the impact of artificial intelligence on the human rights of children;

- encouraging collaboration and interdisciplinary research between AI ethics education and STEM (science, technology, engineering and mathematics) disciplines; • in supporting the development of policies on artificial intelligence and helping to cultivate awareness of the ethics of AI [1].

The pervasive impact that AI has on modern societies and communities leads us to reflect on how certain tools can be used to support inclusive education. In fact, these changes lead to reconsidering the way of learning, communicating and interacting in real and digital spaces. In this sense, the educational institution must recognize the benefits of AI and promote teaching action that is aware of the use of technologies, both by students and teachers.

UNESCO, in a recent document, published in 2023, reflects precisely on the importance of considering AI for the future of fair and inclusive education, warning against the risks and enhancing the potential: *"in the recent past, we could be certain that terms such as 'learning', 'educating', 'training', 'coaching', 'teaching' concerned human beings. This is now less clear. The business of 'educating' and 'training' machines is big, global and growing. It is also increasingly an area of competition, between private companies and actors, as well as nation states. Billions of dollars are now being invested in generative AI companies, when they could be directed towards teacher development and making needed improvements to schools and other physical and social infrastructure that benefit children. It is conceivable that the investments directed to making AI smarter and more capable might someday surpass the investments directed towards educating children and other people"* [2].

Given the revolutionary impact that technologies will have in our lives, it is considered necessary to reflect on the educational implications of artificial intelligence, underlining the strengths. According to a review of the scientific literature carried out by Bouck et al. [3] in 2021 AI could support inclusive education in order to respond to the needs of all students by modifying study programs, offer technological solutions contextualized and adapted to different contexts, promote socio-cultural solutions that develop sensitivity cultural.

According to UNICEF [4], inclusive education is the basis of a more equal and quality society and authors such as Collins and Halverson [5] state that technologies, if supported by an adequate system of knowledge, could offer solutions innovations to improve the educational experience and student learning [6]. Therefore, in wanting to summarize the multiple ways of using AI in the educational context, the personalization of learning is included to adapt the teaching material and activities based on the specific needs of all students. This allows teachers to offer personalized and targeted educational support that takes into account the specific abilities and challenges of each student. It could be an ally in the selection of assistive tools and technologies such as speech recognition software, assisted communication tools, reading devices or text-tospeech software for sensory disabilities. Such tools would allow students to actively participate in learning.

Furthermore, Artificial Intelligence could help teachers accurately and objectively monitor and evaluate student performance, identify areas of strength and improvement, and adapt teaching strategies accordingly.

In summary, integrating artificial intelligence into traditional teaching could support the teacher's teaching action and encourage the active participation of all students, allowing them to reach their full educational potential.

2 The Use of Chatbots in Teaching Research

In an article appearing in The Economist in the United Kingdom and reported by the weekly magazine Internazionale on 6 October 2023, it is specified that "AI applied to research could significantly accelerate the pace of discoveries and forever change our way of understanding knowledge". In support of this thesis, AI scholars Demis Hassabis and Yann LeCun state that in recent years the productivity of scientists, linked to the use of AI, has increased in two sectors: in research in which a literature review is carried out scientific and in research in which new hypotheses are formulated on data already existing in scientific literature. In both cases, the connections that AI reproduces could generate new ideas and develop unexpected theories, free from prejudice [7].

Educational research on AI is still ongoing, but has already produced promising results. Specifically, some research is focusing on the use of chatbots, systems that provide immediate responses and improve student interaction on online platforms [8]. It would seem, in fact, that chatbot technology would offer students a more personalized, engaging and accessible learning environment [9].

Several studies suggest that using chatbot AI in classrooms improves disciplinary skills [10] and interactions between students and teachers [11].

Chatbots versus traditional teaching and learning approaches. In particular, chatbots offer personalized and engaging learning experiences and are capable of adapting disciplinary content to the individual needs of students. This highly personalized approach would improve motivation and learning outcomes [12].

Chatbots offer accessible and flexible learning for students as they are available at any time and in any place, plus they can be used from any device connected to the Internet.

Chatbot technology is able to collect data on students in a short time to improve the effectiveness of learning.

Ultimately, the applications of chatbots in the educational sector are highly flexible and could find application in the form of tutoring activities, such as, for example, the explanation of concepts or the resolution of problems.

According to Google's AI Bard technology, the chatbots most used in teaching are those that provide support to students, both in terms of tutoring and management of teaching activities. Tutoring chatbots are designed to provide students with personalized learning support, such as explaining concepts or solving problems. Some examples of chatbots for tutoring include:

- Socratic is a chatbot developed by Google AI that can help students understand concepts in multiple disciplines such as science, mathematics, literature or social studies. With the support of teachers, this AI offers visual explanations of important concepts in different disciplines [13].
- Marvin is a chatbot developed to offer learning through the use of chats [14].

- Tutorbot is a chatbot developed by the University of Cambridge that can help students improve their language skills [15].
- Bing is a chatbot developed by Microsoft that can be used to search for information on the web, to chat with users on a variety of topics, to create creative content, such as poems, stories or songs [16].
- Bard, is a chatbot created by Google AI that is based on a large language model. With this system it is possible to generate texts, creative contents and informative answers to questions [17]. Bard is still in development, but has the potential to become a valuable tool for education [18].
- ChatGPT [19], launched in 2022 by OpenAI, is a system that allows you to generate text, produce diverse creative content and provide informative answers to questions. Currently the system has two versions: a free one called ChatGPT 3.5 and a paid one called 4.0 [17].

Table 1. Chatbots in comparison in Rudolph et al. p. 376 [17]

Chatbot	Price	Features
ChatGPT (GPT 3.5)	Free	Conversational Code- writing capability More simplicistic and formulatic than GPT-4
ChatGPT (GOT 4.0)	US$20 per month	Largely a more sophisticated version than ChatGPT-3.5 with more precise and articulate prose
Bing Chat	Free	Internet access Provides hyperlinks to sources Uses GPT4
Bard	free	"google" feature conversational

In Table 1, Rudolf et al., list the AI platforms most used by users, not only for educational purposes.

Instead, chatbots used mostly in the management of educational activities are designed to automate administrative tasks, such as collecting assignments or distributing educational materials. Some examples of chatbots for managing educational activities include:

- ClassDojo is a chatbot that can be used by teachers to track student progress [22].
- Schoology is a chatbot that can be used by schools to distribute educational materials and manage student activities [23].
- Google Classroom is a chatbot that can be used by students to access teaching materials and communicate with teachers [24].

It should be noted that the scientific literature in the teaching field is still in its infancy, there do not appear to be any protocols on how to use Artificial Intelligence in the right ways at school, therefore the tools presented must be understood as possible technological allies of teachers. In fact, when adopting Artificial Intelligence, the teacher must necessarily refer to creativity, flexible and simple action which are the key elements of the teaching posture [25].

3   A Possible Use of Chatbots at School for the Promotion of STEM Skills Through the TEAL Model

Technology Enabled Active Learning (TEAL) is an approach to teaching that incorporates technology to improve the student learning experience by promoting active and participatory engagement. This method is based on the strategic use of technology to enhance learning and encourage more active student participation. This innovative teaching method is especially suitable for promoting learning in STEM disciplines.

STEM (Science, Technology, Engineering and Mathematics) skills are a set of key skills and are increasingly in demand in modern society. STEM skills go beyond knowledge of scientific or mathematical facts and include cognitive, social and practical skills.

The model was born around the nineties, at the Massachusetts Institute of Technology (MiT) in Boston, where some researchers, including John Belcher, Peter Dourmashkin and David Litster, created a mix of innovative teaching, technology and classroom design [26].

Specifically, TEAL teaching involves the redefinition of spaces and the structuring of the teaching material that will be delivered to students in advance. In this case the teacher could decide to have his students use chatbots to allow them to have immediate feedback on learning. These are just two of the characteristics that make the model innovative, in fact, another element is the promotion of collaborative learning through Web and Learning Management System (LMS) tools and digital platforms [27]. The teaching method is considered "mixed" as it combines online and face-to-face learning, using both digital and traditional teaching mediators if necessary. The tools necessary to start TeaL teaching are:

- synchronous communication software;
- digital devices;
- sharing of material;
- internet access;
- digital or virtual environments [28].

The fundamental principles of the method are:

- cooperative learning;
- the use of "concept questions", interactive questions with the use of automatic responders;
- immediate feedback on answers from the teacher and automatic responders;
- the use of technologies to communicate, research, observe and present content.

The declination of the TeaL method in the scholastic context can be summarized in four phases:

1. Activation - the teacher provides the students with a topic that captures their attention and activates an interest in learning in them (engagement must be activated). The activity can be sent to students "remotely" and in this case the teacher could suggest that students use chatbots to generate questions and receive simple and immediate answers.
2. Production - the activity takes place in person and allows the students to analyze the issue and the teacher to propose different active teaching strategies, such as: carrying out a project, carrying out an investigation, solving a problem. In this case, the teacher could ask students to try to identify flaws in a chatbot's answers and try to implement the explanation.
3. Elaboration - each group will present the essay through a collective process of reflection and comparison on what has been learned, with the aim of consolidating the learning.
4. Closing - the teacher summarizes what emerged from the groups, focuses attention on the main elements of the skills learned and creates a connection link for the next TeaL lesson [28].

Implementing TEAL requires careful planning, adequate training for teachers and the choice of technologies suited to teaching needs. When well designed, TEAL can help improve the effectiveness of learning and prepare students for modern challenges by considering the strengths of artificial intelligence.

4 Conclusions

In recent years, the use of innovative teaching tools and techniques to support researchers, educators or students has become increasingly frequent in the educational field [23]. This trend seems likely to increase in the future, therefore it is necessary to critically evaluate the ethical and technical implications of AI and ensure that it is used in a responsible and transparent way. In fact, the conscious use of technologies is fundamental to fully realize the potential of AI in research and education [24].

Certainly, the ethical issues related to the use of AI in classrooms raise concerns related to the assessment of skills. In this sense, it is important that the teacher uses chatbots in a constructive way, as a form of support and not as a replacement for human competence, judgment and creativity [19]. The teacher's didactic action could have significant advantages and have positive implications in the planning of teaching activities, in the creation of personalized materials, activities and teaching content for students [30]. In this sense, in order to consider technology as an opportunity it is necessary to develop that sense of co-living with the various AI systems in order to "increase awareness, adopt adequate laws and consolidate ethical values" [30] (p.13).

References

1. UNESCO. Guida per l'Intelligenza Artificiale generativa nell'educazione e nella ricerca, p. 15 (2023). http://unescoblob.blob.core.windows.net/pdf/UploadCKEditor/Brochure%20su%20Raccomandazione%20UNESCO%20sullIntelligenza%20Artificiale.pdf

2. UNESCO, Generative IA and future of the education, p. 7 (2023). https://unesdoc.unesco.org/ark:/48223/pf0000385877.locale=en
3. Bouck, E.C., Flanagan, S.M., Miller, S.J., Bassette, L.: Using artificial intelligence to support the inclusive classroom. Interv. Sch. Clin. **56**(5), 286–293 (2021)
4. UNICEF. Conceptualizing Inclusive Education and Contextualizing it within the UNICEF Mission (2014). https://www.unicef.org/eca/sites/unicef.org.eca/files/IE_Webinar_Booklet_1_0.pdf
5. Collins, A., Halverson, R.: Rethinking education in the age of technology: the digital revolution and schooling in America. Teachers College Press, New York, NY, USA (2018)
6. Bransford, J., Brown, A., Cocking, R.: How people learn: brain, mind, experience, and school. The National Academies Press, Washington, WA, USA (2000)
7. La nuova rivoluzione scientifica. The Economist in Internazionale, n. 1532- anno 30, pp. 45–46 (2023)
8. Perrenoud, P.: dix nouvelles compétences pour enseigner. invitation au voyage. eSf, Paris (trad. it. dieci nuove competenze per insegnare: nvito al viaggio, anicia, roma 2002) (1999)
9. Sibilio, M.: La didattica semplessa. napoli: Liguori (2014)
10. Smutny, P., Schreiberova, P.: Chatbots for learning: a review of educational chatbots for the Facebook Messenger. Comput. Educ. **151**, 103862 (2020)
11. Cunningham-Nelson, S., Boles, W., Trouton, L., Margerison, E.: A review of chatbots in education: Practical steps forward. In: 30th Annual Conference for the Australasian Association for Engineering Education (AAEE 2019): Educators Becoming Agents of Change: Innovate, Integrate, Engineers Australia, Motivate (2019)
12. Durall, E., Kapros, E.: Co-design for a competency self-assessment chatbot and survey in science education. Springer, International conference on human-computer interaction (2020)
13. Mendoza, S., Hernández-León, M., Sánchez-Adame, L.M., Rodríguez, J., Decouchant, D., Meneses-Viveros, A.: Supporting student-teacher interaction through a Chatbot. In: Zaphiris, P., Ioannou, A. (eds.)7 th International Conference, LCT 2020, Held as Part of the 22nd HCI International Conference, HCII 2020, Copenhagen, Denmark, July 19–24, 2020, LNCS, vol. 12206, pp. 93–107. Springer (2020). https://doi.org/10.1007/978-3-030-50506-6
14. Chen, L., Chen, P., Lin, Z.: Artificial intelligence in education: A review. IEEE Access **8**, 75264–75278 (2020)
15. Socratic homepage. https://socratic.org/. Accessed 2 Nov 2023
16. Marvin homepage. https://www.trymarvin.com/. Accessed 2 Nov 2023
17. Tutorbot homepage. https://dictionary.cambridge.org/dictionary/english/chatbot. Accessed 2 Nov 2023
18. Bing homepage. https://www.bing.com/?/ai. Accessed 2 Nov 2023
19. Rudolph, J., Tan, S., Tan, S.: War of the chatbots: Bard, Bing Chat, ChatGPT, Ernie and beyond. The new AI gold rush and its impact on higher education. J. Appl. Learn. Teach. **6**(1) (2023)
20. Bard homepage. https://bard.google.com/. Accessed 2 Nov 2023
21. ChatGPT homepage. https://chat.openai.com/. Accessed 2 Nov 2023
22. Classdojo homepage. https://www.classdojo.com/it-it/. Accessed 2 Nov 2023
23. Schholgy homepage. https://app.schoology.com. Accessed 2 Nov 2023
24. Google Classroom homepage classroom.google.com. Accessed 2 Nov 2023
25. Sibilio, M., Zollo, I.: Corporeità e postura nell'interazione didattica: alcune riflessioni. In: Embodiment & School, pp. 317–323. Pensa Multimedia (2020)
26. TeaL information Website. (2005). http://web.mit.edu/edtech/case-studies/teal.html. Accessed 1 Dec 2023

27. Dourmashkin, P., Tomasik, M., Rayyan, S.: The TeaL physics project at MiT. In: Mintzes, J.J., Walter, E.M. (eds.) Active Learning in College Science: The Case for Evidence-Based Practice, pp. 499–520. Springer International Publishing, Cham (2020). https://doi.org/10.1007/978-3-030-33600-4_31

28. Panzavolta, S., Cinganotto, L.: apprendere le STeM con la me- todologia TeaL. Quando la tecnologia supporta l'apprendimento per problemi. IUL Res. **1**(2), 133–153 (2020)

29. Di Tore, S.: Dal metaverso alla stampa 3D: Prospettive semplesse della didattica innovativa. Studium edizioni, Roma (2022)

30. Kooli, C.: Chatbots in education and research: a critical examination of ethical implications and solutions. Sustainability **15**, 5614 (2023). https://doi.org/10.3390/su15075614

31. Labadze, L., Grigolia, M., Machaidze, L.: Role of AI chatbots in education: systematic literature review. Int. J. Educ. Technol. High. Educ. **20**(1), 1–17 (2023)

Peek the Edu-Metaverse: From an Educational Framework to the AI Challenges for Biometrics

Umberto Bilotti[1]([✉]), Fabrizio Schiavo[2], Pio Alfredo Di Tore[2], and Michele Nappi[1]

[1] Università degli Studi di Salerno, Salerno, Italy
{ubilotti,mnappi}@unisa.it
[2] Università degli Studi di Cassino e del Lazio meridionale, Cassino, Italy
{fabrizio.schiavo,pioalfrado.ditore}@unicas.it
https://www.unisa.it/, https://www.unicas.it/

Abstract. The confirmed interest of the scientific community, the conspicuous investments of the major stakeholders and the reflections that have matured as a result of the extensive use of distance learning seem to suggest that the horizon of a metaverse for the educational context is becoming sharper and closer. Many of the peculiarities of distance education that a few years ago might have seemed ancillary or palliative are now necessary and decisive. One of the greatest efforts made by scientific research in recent years has been to untangle the potential and problems of Edu-Metaverse from the tangle resulting from a structural formulation that contemplates advanced and interconnected technologies. In the first part of the paper, the causes that require the development of a new educational-didactic model are taken up. Then, some possible technology-learning theory pairings are identified to support the Edu-Metaverse. Finally, an attempt will be made to verticalise by identifying possible challenges for Artificial Intelligence developments in Biometrics.

Keywords: Education · Metaverse · Artificial Intelligence · Biometrics

1 Introduction - Toward a New Educational Model

Under the assumption that educational action is the implementation of useful strategies to foster the learning environment, a predictive view of the learner's context of interest becomes indispensable. The first educational system based on class composition, *the Prussian model*, was introduced as a response to the needs arising from the first industrial revolution. In fact, the one-to-many teacher-student scheme ensured a sufficient level of knowledge and skills for the new, evolving society. Today, the traditional education system risks producing citizens detached from their personalities and asynchronous with today's cultural rhythms. The global spread of the computer, the advent of the Internet has been crucial for the expansion of the learning environment from offline to online space [18]. The possibilities offered by a large number of didactic softwares and the usability of the current Learning Management Systems (LMS) were largely experienced by the native digital generation, but several issues and limitations emerged not only from the technologies but also from their use that was not supported by a consistent pedagogical framework. Several studies such as the work of R. Yavich and B. Starichenko [26] have shown that the educational methods for a virtual environmented must have certain structural-relational characteristics. However, the proposed design can be a new starting point for a new educational model that can consider a larger set of technologies and learning theories. In the Sect. 2, some possible technology-learning theory pairings consistent with a new educational framework are described while Sect. 3 isolates two of the main effective components for the introduction of the learner in the Edu-Metaverse, Sect. 4 focuses on potentials and challenges of AI methods in the biometric domain, and finally in Sect. 5 conclusions are given.

2   The Proposal of an Educational Metaverse

Although the concept of the metaverse could contemplate many different applications, the literature has managed to identify certain characteristics that, if pursued, become guidelines for the realisation of a new digital environment. If we define a metaverse as a *shared, immersive, and 3D computer-generated* universe, many of the technologies required to create such an environment are already extensively utilized and their adaptability could extend the pool of applications to the educational context [5]. Below, but without claiming to be exhaustive, we will analyse three of the possible pairings of metaverse technology and learning theory that are useful in justifying the effort required of the scientific community for the realisation of an Educational Metaverse.

2.1   Blockchain and Peer-to-Peer Learning

The implementation of a distributed public ledger for data management, such as that of a blockchain, is one of the main strategies for a shared metaverse. Many of the blockchain-based structures for the educational environment that have been developed are not limited to high-level issues such as the secure migration of sensitive data and certifications between university and corporate consortia, but others are intended as their own support for teaching. One of the most significant examples is the work of Duong, T. N. B., and Jun, J. Y. T. [10] in which LMS is realised exploiting a blockchain such as Ethereum. EtherLearn is an LMS whose management is therefore non-centralised and equipped with a reward system for information uploaded by a specific user, weighted according to the quality of the information assessed by other users. The choice of an LMS such as Etherlearn is certainly appropriate for a teaching strategy that includes a peer-to-peer component. The peer-to-peer learning model in fact dispenses with the hierarchical teacher-student structure, in which contributions to solving a problem are condensed into the knowledge emanating from the teacher, and instead promotes a network of intelligences in which contributions to solving a problem can potentially be uniform. In particular, in addition to incentivizing the production of original material by the students, the quality of the content is representative of the class that validates it, being, for the teacher, a parameter of the general level of understanding.

2.2 Extended Reality and Symplex Didactics

The first Augmented Reality (AR) and Virtual Reality (VR) technologies can be traced back to the 1950s and 1960s, but their use for a wide audience and in the educational-didactic context has been studied in recent decades. The study of Hincapie et al. [14] on AR applications for educational-didactic purposes reveal that content meant to 'augment reality,' such as 3D models and animations, has a greater influence on memory and motivation than other 2D digital information, such as text, photographs, and videos. Also in the work of D. Hamilton et al. [11] VR applications of this type are analysed and all associated tests confirm the hypothesis that the improvement in learning is directly proportional to the level of immersiveness of the proposed VR technology. From an educational and didactic point of view, the implementation of these particular technologies has several implications. However, the main consequence derives from a new high degree of motor involvement required during the use of immersive devices. The crucial aspect of the motor component during the human learning process is certainly at the centre of the simplex didactics proposal [23], a declination of the homonymous theory proposed by A. Berthoz [3]. Indeed, human learning, as a strategy of adaptation to the environment, requires multiple and diverse interactions with the latter. Consequently, for our body, as a knowledge machine, the construction of meaning occurs more efficiently when it is aided by natural actions such as the movements of our body encoded then as commands from our avatar.

2.3 Digital Twin and Seemlessly Learning

A Digital Twin (DT) is a digital representation designed to emulate physical and functional characteristics of a real object or system. Although it was first formulated for the manufacturing or aerospace sector, a variety of other application contexts have been identified over the past decade [20]. One of the most intriguing and challenging hypotheses is the possibility of a learner's DT, as it should be able to approximate some of our cognitive processes. The feedback and feedforward relations between a Cognitive DT (CDT) and its real counterpart should tend towards the non-adversarial symbiote type. In this view, the Symbiotic CDT (SCDT) must be responsive to both the evolution and degradation of its own abilities and knowledge. A further extension of the SCDT can result from the integration of memiotic aspects. The resulting memetic SCDT or memion will then be able to modulate learning dynamically and coordinate with other memions or an entire community [15]. The characteristics thus hypothesised of the future learner's Digital Twin are certainly akin to seamless learning.

This learning style considers the possibility that learning can take place continuously, i.e. without interruptions caused by the transition from one context to another [25]. The Digital Twin can therefore already be effective in following this first direction by lending itself as an adapter between an increasingly diversified demand for competences and the indispensable condition of personalised learning. Seamless learning, however, should not only be understood as potential learning everywhere and everytime, the use of which could already be realised by a sterile on-line didactics, but also and especially in an enactive formula therefore coordinated with the environment in which it takes place. Once again, the Digital Twin can accommodate this suggestion through the reprocessing of large amounts of data from devices installed in the environment or worn by the student himself. The space in which the subject is learning is less and less a constraint rather an opportunity where, starting from the raw data and the right model capable of interpreting them, it is possible to increase the level of didactic awareness, to adapt and improve the learning experiences that the various life contexts can present to the individual subject, integrating them in the best possible way with the processes and paths of a classroom education [19] (Fig. 1).

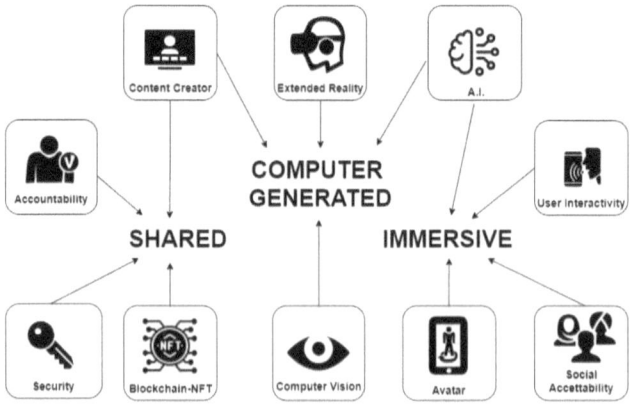

Fig. 1. Some of the main components of metaverse

3   Learner Uploading into the Edu-Metaverse

3.1   The Role of Artificial Intelligence

Following the observations made on the potential offered by an architecture capable of realising the binomials described, it is necessary to identify the relevant open challenges that may separate us from the final product. Certainly, a strong push forward is achieved by the great availability of data and the exponential growth of Artificial Intelligence-based methods. As can be seen from the paper of Zhai et al. [28], in recent years there has been a great deal of research that has experimented with Artificial Intelligence (AI) in the educational teaching context. From their article, three main directions of work on this topic are identified and defined, in particular within the *Dimension of Application*, Artificial Intelligence methods are implemented for the development of affection computing, role-playing, gamification, immersive learning and gamification, those that will in fact become key components of the Edu-Metaverse. Already from the work of He and Niu [12], it is observed that it is necessary to go beyond the primary objective of creating an intelligent digital teacher capable of proposing learning paths tailored to each different student. The hypothesis put forward for the teaching of physical education envisages the creation of an intelligent digital body capable of emulating the student's respective body through the processing of information sensed by wearable devices. This hypothesis, which less than three years ago might have been strongly localised, with the gradual spread of immersive technologies and the advent of various proto-metaverses, is now a necessary condition.

3.2   The Role of Biometrics

Users' physical movements are converted into virtual world activities, allowing them total control over how their avatars interact with other metaverse items. Furthermore, these avatars may interact with a variety of real-world modalities, such as voice recognition and dynamic analysis, which are aided by AI in terms of accuracy and processing speed. Facial expressions, emotions, body's movements, and physical interactions are only some examples of these modalities [5]. These physical and behavioural characteristics, once they become digitalized information, in fact constitute the object of the study of biometrics. This field of computer science, like many others, has from the very beginning been able to exploit AI-based methods and algorithms, thus favouring scientific production capable of solving problems that cut across multiple contexts. Having thus highlighted the fundamental role of biometrics in the development of this framework, we can then focus on the main challenges and possibilities.

4   Biometrics for Edu-Metaverse: AI Solutions

4.1   Cheating Prevenction and Cheating Detection

One of the significant challenges associated with online learning lies in assessing students, as there is a heightened risk of academic integrity violations due to various forms of cheating during online examinations. A study led by F. Noorbehbahani et al. [21] provides a comprehensive perspective on the most challenging issues within online learning, particularly concerning strategies for mitigating, detecting, and preventing cheating. From their research, two main modes of cheating were classified, individual cheating and group cheating, and from both many of the illegal behaviours can be turned into biometric problems. A new possible partition of problems that we can address to biometrics consists of the following three classes: individual cheating using forbidden materials, group cheating trought impersonation and group cheating trought collaboration. The use of forbidden materials such as reading texts, notes or other monitors can, for example, be countered by many of the popular eye-tracking algorithms [7], many of which form the main component of the most popular VR and AR viewers. The issue of impersonation, on the other hand, is more varied as it often depends directly on the mode chosen for the examination. Indeed, impersonation can easily occur in the case where the test consists of keyboard text composition when the microphone and camera are not required to remain active. However, even the very activation of the microphone and video camera can be circumvented through voice conversion in the first case and face presentation or face impersonation in the second case. These types of attacks can instead be responded to through modern Face Recognition, Voice Recognition or Keystroke Dynamic Identification [4] techniques. Many of these solutions can also be adopted in cases of not allowed collaboration or communication, in particular in the identification of sources of information from other persons present in the home environment indicated as the place of examination. An institution or teacher who chooses to equip online examinations with such software is aware that creating a climate that discourages cheating not only ensures a more honest assessment, but can also improve the quality of learning.

4.2   Performance-Enhancing Tools

The effort in the realisation of tools to improve a student's performance should not only end at the stage of assessment of preparation but be directed at every stage of knowledge acquisition. From this consideration, it follows that such tools can also lead to improvement through direct and strongly focused intervention. All those methods that study some of the components of the learning process, such as the attentional and emotional component, can therefore be candidates as such tools.

Level Attention Estimation. Although much of the research in the field of Level Attention Estimation has utilised technologies such as electroencephalogram, electrocardiograms and electromyogram, since the 1970s, several studies have been conducted to investigate the relationship between eye blink rates and a person's cognitive activity, such as their degree of attention [2]. Several works have demonstrated the effectiveness of eyelid blinking as a means of detecting drowsiness, in particular through the analysis of factors such as frequency, duration, eyelid speed and percentage of eye closure. In addition to these characters, other biometrics that can be used as indicators of drowsiness may be variations in yawning frequency and abnormalities in pupil dilation [6]. The relations found were used to study attentional performance during online courses by Daza et al. [8]. Unlike the presence or absence of sleepiness, the attentional state cannot be investigated as a binary event; rather, it should be seen more broadly as a continuous phenomenon, even though the data to be evaluated may be the same. The determination of a blink is the goal of the Convolutional Neural Network they proposed and after formalising the relationship between attention level and blink frequency, a classification of three attention levels (high, medium and low) resulted. The knowledge of this information in real time, can be a useful tool for the teacher, both for a targeted choice of interactions and for the teaching design itself. However, as the authors themselves envisage, to improve accuracy it would be advisable to implement a method that also analyses the different human-machine interactions depending on the devices used, so from the keyboard and mouse of the computer to the touchscreen of smartphones to joypads or one's own body detected by VR and AR visors.

Facial Emotion Recognition. Even in a more generic context than the educational-didactic one, the level of attention is strongly linked to a person's emotional state. Apart from a few particular cases, the human being, even if not voluntarily, is capable of expressing his or her emotional state through various non-verbal communication channels, the one with the greatest capacity for information certainly being the face. Facial Expression Recognition is the branch of Biometrics that proposes and develops methods capable of classifying the main human emotions from information gathered from images, audio, video etc. At present, many of the AI-based methods have achieved high levels of performance in terms of accuracy and calculation time, but a number of relevant questions remain open. The first inherent to the very nature of emotion as it is considered a dynamic phenomenon and whose detection should therefore result from a video or audio file that unfolds over an interval of time. The second relates to the necessary univocal emotion-file correspondence of a manifestation that is instead "fuzzy" because it is produced by varied cultural, social and psychic components. According to research, the most discriminating areas of the face surfaces that contribute to facial emotion detection are situated on the mouth and the eyes. Furthermore, [17] investigated the influence of different facial regions to different emotions. Their findings indicate that the mouth has the best categorization accuracy since it appears to hold a lot of information regarding facial expression. The mouth, in particular, outperforms the other main facial parts such as the nose and eyes for the Neutral, Happiness, and Anger emotions. Instead, the single emotion in which eyes cause the least confusion is fear. The experimental results provided by Abate et al. [1] reveal an expected loss of accuracy in all circumstances that introduce an occlusion, compared to the level of performance achieved with the face fully visible. On the other hand, occlusions of the periocular regions are not as severe as those caused by face masks. We can therefore hypothesise that it will be possible to realise accurate methods of RES that are resistant to occlusion caused by wearing VR and AR visors and ultimately detect not only states such as anger or sadness that are not conducive to learning, but also states such as disgust as a symptom or consequence of physical discomfort caused by VR sickness.

4.3 New Inclusive Technologies

The considerations that have been made in the preceding paragraphs have been useful in a direct way, to identify certain methods, technologies or entire subsectors of Biometrics research useful for the chosen cause. However, in an indirect way, they will allow us to contemplate a last but increasingly established perspective of applications such as those of inclusive technologies. As tools such as hardware, software, and peripherals designed with the aim of fostering the process of inclusion, especially within the classroom context, the main target group for which these tools are intended are certainly students with special educational needs, especially students with learning disorders. In these cases, the tension towards customised teaching is greater, as it is no longer a desirable condition for high performance but often an indispensable condition for satisfactory personal development. Several of these tools are described in the work of Yenduri et al. [27] and the versatility of AI emerges from the variety of case studies treated. One of the technologies that has long been used in this field is certainly eye-tracking. Already in the 1980s and 1990s, several studies were able to identify abnormalities in both the visual trajectories and fixation times of dyslexic readers compared to normal readers [22]. Failure to maintain the visual spotlight on specific visual areas for long enough to correctly process the visual stimulus is a deficiency that spills over into the development reading competence, making it difficult to identify letters and words [9]. Unfortunately, while for diagnosis eye tracking is a universally suitable tool (except for visise disabilities), for support it will never be sufficient or definitive. For example, one of the most popular assistive technologies, used not only by dyslexic learners, such as end-to-end speech recognition software will be able to solve problems such as multi-language input, speaker adaptation and unification of speech separation, signal processing and speaker identification modules with lower computational costs [16]. Tan et al. [24] created an evaluation technique for ADHD, a prevalent neurodevelopmental disease marked by hyperactivity, inattention, and impulsivity. They conducted an experiment using multiple sensors and VR technologies and emulated a classroom environment in which continuous performance activities and the Wisconsin card sorting exam were used to assess students' attention, abstract thinking skills, and cognitive competence. Finally they paired their test findings with physiological data such as head and eye movements and EEG to measure the subject's sustained attention and attention shift [24]. Another interesting work is by Herrero and Lorenzo [13] in which an immersive 3D virtual environment is created to improve the social and emotional skills of high-functioning in Autism Spectrum Disorder (ASD) students. Students, after wearing a 3D visor, can interact with human avatars and in an environment both designed for their comfort. The improvements recorded by students with ASD come from being able to practise with a structured but static set of iterations. The use of AI could create avatars capable of generating different iterations while preserving the canons established in the first version. Finally, a further improvement might arise from providing the virtual environment with a FER metod, which is capable of analysing the ASD student's emulation of a specific required emotion.

5   Conclusion

The use of a technology to foster the learning process should not be understood as a strategy intended to trivialise a goal, and this is certainly a risk that predates the advent of AI. The suggestions and ideas put forward in this paper confirm the crucial role of AI in the realisation of new technologies for education and suggest its declination in the biometric sphere in order to foster the development of the Edu-Metaverse in a way that is in line with and coherent with the capabilities of the learner.

References

1. Abate, A.F., Cimmino, L., Mocanu, B.C., Narducci, F., Pop, F.: The limitations for expression recognition in computer vision introduced by facial masks. Multimed. Tools Appl. **82**(8), 11305–11319 (2023)
2. Bagley, J., Manelis, L.: Effect of awareness on an indicator of cognitive load. Percept. Mot. Skills **49**(2), 591–594 (1979). https://doi.org/10.2466/pms.1979.49.2.591
3. Berthoz, A.: Simplexity simplifying principles for a complex world. Hors collection (2019)
4. Bilotti, U., Bisogni, C., Castiglione, A., Nappi, M., Pero, C.: User identification through hidden markov model-based touch keystroke dynamics. In: 2023 5th International Conference on Bio-engineering for Smart Technologies (BioSMART), pp. 1–4. IEEE (2023)
5. Bilotti, U., Di Dario, D., Palomba, F., Gravino, C., Sibilio, M.: Machine learning for educational metaverse: How far are we? In: 2023 IEEE International Conference on Consumer Electronics (ICCE), pp. 01–02. IEEE (2023)
6. Bisogni, C., Hao, F., Loia, V., Narducci, F.: Drowsiness detection in the era of industry 4.0: are we ready? IEEE Trans. Ind. Inf. **18**(12), 9083–9091 (2022)
7. Casanova, A., Cascone, L., Castiglione, A., Nappi, M., Pero, C.: Eye-movement and touch dynamics: a proposed approach for activity recognition of a web user. In: 2019 15th International Conference on Signal-Image Technology & Internet-Based Systems (SITIS), pp. 719–724 (2019). https://doi.org/10.1109/SITIS.2019.00117
8. Daza, R., DeAlcala, D., Morales, A., Tolosana, R., Cobos, R., Fierrez, J.: Alebk: Feasibility study of attention level estimation via blink detection applied to e-learning. arXiv preprint arXiv:2112.09165 (2021)
9. Di Tore, S.: La tecnologia della parola. Didattica inclusiva e lettura, Franco, Angeli (2016)
10. Duong, T.N.B., Jun, J.Y.T.: Etherlearn: decentralizing learning via blockchain. In: 2021 IEEE International Conference on Engineering, Technology & Education (TALE), pp. 212–217. IEEE (2021)
11. Hamilton, D., McKechnie, J., Edgerton, E., Wilson, C.: Immersive virtual reality as a pedagogical tool in education: a systematic literature review of quantitative learning outcomes and experimental design. J. Comput. Educ. **8**(1), 1–32 (2021)
12. He, Z., Niu, X.: Applying artificial intelligence to primary and secondary school physical education. In: 2021 2nd International Conference on Information Science and Education (ICISE-IE), pp. 1577–1581 (2021). https://doi.org/10.1109/ICISE-IE53922.2021.00349

13. Herrero, J.F., Lorenzo, G.: An immersive virtual reality educational intervention on people with autism spectrum disorders (ASD) for the development of communication skills and problem solving. Educ. Inf. Technol. **25**, 1689–1722 (2020)
14. Hincapie, M., Diaz, C., Valencia, A., Contero, M., Güemes-Castorena, D.: Educational applications of augmented reality: a bibliometric study. Comput. Electr. Eng. **93**, 107289 (2021)
15. Kinsner, W.: Digital twins for personalized education and lifelong learning. In: 2021 IEEE Canadian Conference on Electrical and Computer Engineering (CCECE), pp. 1–6. IEEE (2021)
16. Li, J., et al.: Recent advances in end-to-end automatic speech recognition. APSIPA Trans. Sign. Inf. Process. **11**(1) (2022)
17. Lian, Z., Li, Y., Tao, J.H., Huang, J., Niu, M.Y.: Expression analysis based on face regions in real-world conditions. Int. J. Autom. Comput. **17**, 96–107 (2020)
18. Lin, H., Wan, S., Gan, W., Chen, J., Chao, H.C.: Metaverse in education: Vision, opportunities, and challenges. In: 2022 IEEE International Conference on Big Data (Big Data), pp. 2857–2866. IEEE (2022)
19. Mangione, G.R., Di Tore, P.A., Di Tore, S., Corona, F.: Educare seamlessly. dalla visione integrata delle teorie alle esperienze della comunità pedagogica italiana. Italian J. Educ. Res. **14**, 35–48 (2015)
20. Mihai, S., et al.: Digital twins: a survey on enabling technologies, challenges, trends and future prospects. IEEE Commun. Surv. Tutorials **24**(4), 2255–2291 (2022). https://doi.org/10.1109/COMST.2022.3208773
21. Noorbehbahani, F., Mohammadi, A., Aminazadeh, M.: A systematic review of research on cheating in online exams from 2010 to 2021. Educ. Inf. Technol. **27**(6), 8413–8460 (2022)
22. Rello, L., Ballesteros, M.: Detecting readers with dyslexia using machine learning with eye tracking measures. In: Proceedings of the 12th International Web for All Conference, pp. 1–8 (2015)
23. Sibilio, M.: Simplex didactics: a non-linear trajectory for research in education. Rev. Synth. **136**(3–4), 477–493 (2015). https://doi.org/10.1007/s11873-015-0284-4
24. Tan, Y., et al.: Virtual classroom: An adhd assessment and diagnosis system based on virtual reality. In: 2019 IEEE International Conference on Industrial Cyber Physical Systems (ICPS), pp. 203–208. IEEE (2019)
25. Wong, L.H.: A learner-centric view of mobile seamless learning. Br. J. Edu. Technol. **43**(1), E19–E23 (2012)
26. Yavich, R., Starichenko, B.: Design of education methods in a virtual environment. J. Educ. Train. Stud. **5**(9), 176–186 (2017)
27. Yenduri, G., et al.: From assistive technologies to metaverse–technologies in inclusive higher education for students with specific learning difficulties: a review. IEEE Access (2023)
28. Zhai, X., et al.: A review of artificial intelligence (AI) in education from 2010 to 2020. Complexity **2021**, 1–18 (2021)

Breaking Barriers in the Metaverse: A Comprehensive Exploration of Accessibility for Users with Disabilities

Dario Di Dario$^{(\boxtimes)}$ ⓘ, Giulia Sellitto ⓘ, Viviana Pentangelo ⓘ,
Maria Lucia Fede ⓘ, and Filomena Ferrucci ⓘ

University of Salerno, Fisciano, Italy
ddidario@unisa.it

Abstract. In the post-pandemic era, more and more people are turning to virtual spaces for entertainment, work, education, and social interaction. Despite this enthusiasm, serious concerns have arisen regarding the accessibility and inclusion of people with disabilities in educational settings and how Artificial Intelligence (AI) approaches can mitigate these issues in the new digital realm. In this article, we conduct a systematic literature review (SRL) to identify the set of challenges that people and students with disabilities encounter when they are interested in accessing the metaverse. We collect solutions proposed in the literature to provide a comprehensive understanding of the accessibility of current metaverse platforms. We distill a number of take-away messages encouraging researchers and practitioners to further work on the proposed solution, leveraging AI to monitor and enable people with disabilities to fully enjoy the metaverse experience.

Keywords: Metaverse · Accessibility · Disabilities · Systematic Literature Review

1 Introduction

The metaverse has emerged as a promising domain for enhancing human engagement experiences in modern society, characterized by deep interconnectivity and digital integration. This three-dimensional virtual environment provides a large plethora of possibilities across entertainment, education, work, and social interaction, redefining conventional perceptions of digital reality. The educational setting appears to be a valuable scenario for leveraging the metaverse, as it could offer a digital environment enhanced by analytical tools and technologies capable of monitoring not only the academic progress, but also the social and psychological situation of students. For example, previous work has argued that Artificial Intelligence (AI), which is one of the main driver technologies of the metaverse, could play a crucial role in reducing accessibility barriers for students with disabilities. [3] However, despite the hype it is currently subject to, the metaverse still presents some issues; in fact, this paradigm shift has shed

F. Palomba and C. Gravino (Eds.): WAILS 2024, LNCS 14545, pp. 45–55, 2024.
https://doi.org/10.1007/978-3-031-57402-3_6

light on the accessibility and inclusivity challenges faced by individuals with disabilities in educational environment [14], highlighting the need for customized AI tools to address inclusivity challenges [3].

In this paper, we aim to identify the set of challenges that people with disabilities face when trying to access the metaverse, with the corresponding solutions currently available and envisioned at the state of the art. To do this, we perform a Systematic Literature Review (SLR) to gather information on the current landscape of accessibility in the metaverse, ultimately distilling practical take-away messages that we believe can drive further research. Moreover, our finding could leverage AI tools, enhancing the overall metaverse experience for everyone.

2   Background

The term "metaverse" was initially introduced in Neal Stephenson's science fiction novel *Snow Crash*, which was published in 1992 [22]. Nowadays, what began as a fictional concept has since evolved into a tangible reality. Various interpretations of the metaverse have emerged in different application domains, but several key features can generally characterize it. Concretely, the metaverse can be defined as a three-dimensional online environment where users, represented by avatars, engage with each other in real-time within virtual spaces, free from the physical limitations of the real world [20].

One of the application contexts where research focuses the most on the Metaverse is the educational context, exploring its uses in the field of learning science. In fact, following the outbreak of the COVID-19 pandemic, which inevitably shifted attention from in-person to remote learning [2], the Metaverse has been identified as a potential platform capable of overcoming the limitations of traditional 2D platforms [15]. The increased interest in the field of learning science has led to proposals in the literature for frameworks on how to adapt a highly immersive 3D platform such as the Metaverse [1] to various educational contexts and to begin providing functional prototypes developed and tested in educational and academic settings [6, 19].

Since similar platforms are beginning to be utilized in various contexts to overcome the limitations of traditionally known 2D platforms, it is crucial to focus on the types of technologies predominantly used for this purpose and the accessibility and inclusivity challenges they present. As it is a digital world that must be built upon a highly immersive and realistic experience, Virtual Reality technologies are often key enablers for the creation of the metaverse today [14]. Virtual Reality refers to all technologies that enable a completely digitally created world. The creation of such a digital experience that engages all the senses, from sight to hearing and even touch, is often achieved through the combination of specialized multisensory hardware [17]. Currently, the typical main components that work together to make up VR systems include:(1) wearable headsets that deliver both visual and auditory information and incorporate additional sensors to enable input—such as motion tracking through motion sensors and hand tracking via built-in cameras, (2) two handheld controllers for gesture-based input—unless the system supports natural tracking of arms, hands, and fingers—, as well as button-based input, and (3) positional tracking devices, often based on optical technologies, that determine a person's position and orientation within an interaction space. By integrating one or more of these devices simultaneously, implementing metaverse experiences can significantly benefit when considering the level of realism and user engagement [8].

However, when discussing the various devices that enable increasingly immersive interactions in digital platforms, particular attention should be given to the theme of disability. Primarily focusing on the educational context, given the need to provide everyone with equal access to a complete and engaging learning experience, it is necessary to pay attention to the current limitations in terms of accessibility and inclusivity of these technologies and how they could potentially be overcome. To date, the design of digital worlds, interactions, and devices still insufficiently considers people with disabilities. Quinlan [18] highlighted the two primary considerations that should be made when analyzing the exclusion of people with disabilities in digital immersive experiences.

Physical Exclusion. Current devices and interaction methods do not always consider the accessibility needs of individuals with physical disabilities. As an example, VR technologies often prescribe specific actions expected from users, like traditional controls such as button presses or gestures. These actions require precise physical movements that the technology can detect, as seen with sensor-based cameras that necessitate pointing in the air and locomotion. As a result, these interactions may pose significant challenges for individuals with certain disabilities and often lack alternative modes to enhance their accessibility. Therefore, it's crucial to focus on how disability is addressed in systems like these and how technologies can evolve to become more inclusive for everyone.

Social Exclusion. The type of exclusion that may affect people with disabilities extends beyond the physical realm, and social context must also be considered in terms of representation and acceptance. Even though immersive technologies aim to become an integral part of daily life, they often do not adequately acknowledge the needs of individuals with disabilities, further rendering them invisible in the virtual society. Social disparity is evident on two levels: (1) at the level of representation, where often, among the avatar customization choices, people with disabilities find limited options to feel adequately represented by their avatars [24], and (2) at the level of data collection and analysis, where biases are often introduced into information related to disability. For example, Artificial Intelligence (AI) language classifiers may more frequently categorize texts containing terms associated with disability identity as toxic and negative [23].

For these reasons, the theme of disability in metaverse digital platforms is a topic that demands further attention from the research community. Our work aims to systematically analyze existing literature on the subject, gain an overview of the current situation, and extract practical insights to make the metaverse experience more accessible and inclusive.

3   Research Design

This study *aims* to identify the set of challenges that people with disabilities face when trying to access the metaverse with the corresponding solutions currently available and envisioned at the state of the art, with the *purpose* to highlight the difficulties that people with disabilities experience while accessing the metaverse. The *perspective* is from both researchers and practitioners: the former are interested in understanding whether the metaverse can support people with disabilities, and which measures could be taken into account to increase physical and social inclusion; the latter are interested in finding potential solutions to make the metaverse an inclusive place, enabling easy use of it. Furthermore, we believe that the results of this study may offer valuable insights for AI-powered monitoring and support of users in the metaverse.

Given our goal, we performed a Systematic Literature Review, i.e., a meticulous approach that involves various steps to gather, analyze, and report knowledge about the subject under examination. Following the guidelines proposed by Kitchenham [12], we formulated two research questions driving our work:

> **Q RQ$_1$.** *What are the main challenges faced by individuals with disabilities when trying to access the metaverse?*

The primary objective of **RQ$_1$** is to gain insight into the barriers people with disabilities encounter in the metaverse. This knowledge is essential to empower them, influence policy changes, and advance technology for more comprehensive virtual interactions. Furthermore, it raises public consciousness about the importance of digital accessibility.

> **Q RQ$_2$.** *What measures can be taken to enhance accessibility for individuals with disabilities, enabling their active engagement in the metaverse?*

With our **RQ$_2$**, we aimed to explore effective measures that an educational metaverse could implement to address the accessibility challenges identified. It aimed to offer meaningful perspectives on how individuals with disabilities can be integrated into this cutting-edge technology.

The main steps we employed to tackle the RQs mentioned earlier are outlined in Fig. 1, where we depicted the various stages of the SLR process using grey round squares, which are elaborated upon in the subsequent section. We indicated the authors responsible for each phase using green circles. Additionally, we used green rhombuses to denote the authors who reviewed the completed activity. More information was provided in the online appendix of this paper [5]. As a final remark, we report the results by following the *"Systematic Review"* guidelines depicted in the *ACM/SIGSOFT Empirical Standards*[1].

3.1   Search Strategy

Initially, we meticulously analyzed and selected the most relevant research domains encompassing primary and secondary studies, including but not lim-

---

[1] https://github.com/acmsigsoft/EmpiricalStandards.

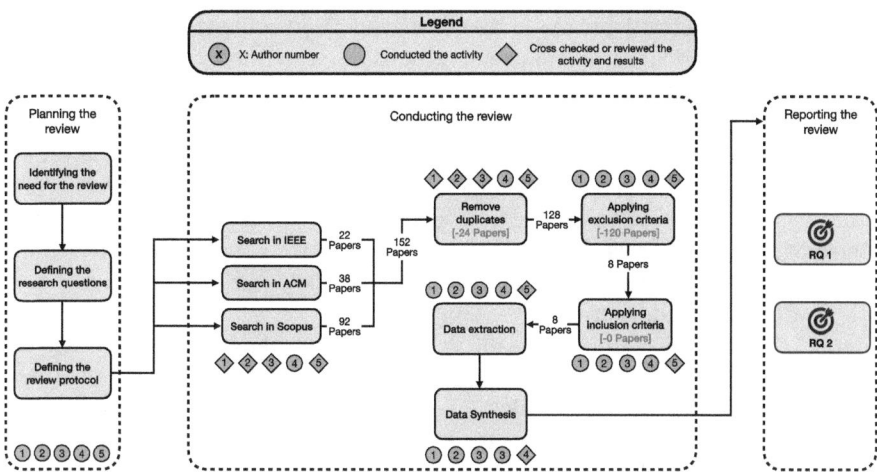

Fig. 1. Research Process Overview.

ited to *Metaverse* and *Disabilities*. It is important to note that our research only focused on the topic of metaverse, excluding virtual reality due to the large number of results. This choice allows for a more focused accessibility examination within the unique and evolving metaverse concept.

After expanded the scope of the study, we enlarged our search by formulating a comprehensive list of synonyms to capture articles on the same topic but with other spellings. To develop a well-defined research query, we used the AND operator to combine the mentioned domains and the OR operator to combine the synonyms. As a result of this phase, we generated the following query:

🔍 **Query.** *"Metaverse" AND ("disability" OR "physical disability" OR "disabled people" OR "accessibility" OR "inclusion")*

The query was run on June 14th, 2023 against the three of the most known research engine, namely *IEEE Xplore*[2], *Scopus*[3], and *ACM Digital Library*[4], obtaining 152 primary studies.

3.2 Article Selection Process

After acquiring articles, we conducted an additional analysis to remove 24 duplicate entries that were identified from multiple data sources. To confirm the validity of our study, we strictly applied the exclusion and inclusion criteria outlined by Kitchenham et al. [12]. These criteria were designed to ensure that only the relevant articles addressing the research questions were considered. We removed

---

[2] https://ieeexplore.ieee.org.
[3] https://www.scopus.com.
[4] https://dl.acm.org.

any article that met at least one of the following exclusion criteria: (1) not peer-reviewed, (2) not related to metaverse accessibility, (3) not written in English, and (4) full text not available for free.

At the end of this phase, 120 papers were discarded, and the remaining 8 papers were considered suitable for further evaluation under the inclusion criteria. The papers that fulfilled at least one of the following inclusion criteria were considered eligible and moved on to the next evaluation phase: (1) articles describing challenges related to accessing the metaverse from people with disabilities, and (2) articles reporting measures to improve the inclusion of people with disabilities in the metaverse.

Following this phase, we have integrated all the examined articles into our study. However, when the number of available studies is limited, conducting a quality assessment may not provide significant insights, which is the case here. As a result, we decided to analyze the available content directly instead of engaging in a potentially less informative quality assessment. For this reason, the authors defined a data extraction form and read the primary studies together to gather relevant information to address the research questions. The data extraction process was constantly monitored by a senior researcher. After this, the last step regards the data synthesis of the extracted data, applying the *thematic analysis* [4] method to analyze and identify common themes arising from the text. The process followed an iterative approach, where each step aimed to refine previously emerged themes. Also in this case, the process was monitored by the senior researcher.

4   Results

Table 1 provides an overview of the eight papers which we finally considered in the Data Extraction process, pointing out the research question(s) they allowed us to answer. In the following, we elaborate on the relevant data extracted from the analysed literature, providing insights on the state of the art on metaverse accessibility, and distil practical take-away messages that we believe can be meaningful for the communities of researchers and practitioners on the matter.

4.1   RQ$_1$: On the Challenges Faced by People with Disabilities When Trying to Access the Metaverse

Our first research question drove us to assess what are the main challenges faced by individuals with disabilities when trying to access the metaverse. We extracted data from three primary studies [8, 10, 21], gathering information on the current limitations of existing metaverse platforms, which lead to accessibility challenges for people with disabilities. We identified five main families of challenges, both related to the intrinsic immersivity of the metaverse, and coming from a suboptimal implementation of the platform.

Lack of Alternative Textual Descriptions. A significant accessibility barrier in the current implementations of the metaverse consists of the lack of alternative

Table 1. Papers participating in the Data Extraction Process.

Ref	Authors	Title	RQ$_1$	RQ$_2$
[10]	Choi and Ka	Empirical Evaluation of Metaverse Accessibility for People Who Use Alternative Input/Output Methods	☑	
[7]	Dudley et al.	Inclusive Immersion: A Review of Efforts to Improve Accessibility in Virtual Reality, Augmented Reality and the Metaverse		☑
[8]	Gerling and Spiel	A Critical Examination of Virtual Reality Technology in the Context of the Minority Body	☑	
[9]	Hadi Mogavi et al.	Envisioning an Inclusive Metaverse: Student Perspectives on Accessible and Empowering Metaverse-Enabled Learning		☑
[11]	Kanza et al.	Geospatial Accessibility and Inclusion by Combining Contextual Filters, the Metaverse and Ambient Intelligence		☑
[13]	Meena et al.	Advancing Education through Metaverse: Components, Applications, Challenges, Case Studies and Open Issues		☑
[16]	Parker et al.	Towards an Inclusive and Accessible Metaverse		☑
[21]	Seigneur and Choukou	How Should Metaverse Augment Humans with Disabilities?	☑	☑

textual descriptions for content and functionalities. Such limitation undermines comprehension for individuals with visual impairments. In fact, such individuals often depend on assistive technologies to navigate digital spaces, and without comprehensive descriptions, they are left at a disadvantage, and are prevented from enjoying the metaverse experience.

Lack of Appropriate Markups. As in the real world people with disabilities rely on markups to orient themselves and perceive the space they are living in, proper markups play a fundamental role in enabling assistive technologies to operate effectively within the metaverse as well. Such markups are essential for people with disabilities to navigate and interact with the virtual world; without them, they face significant challenges in orienting themselves and engaging meaningfully in the digital realm.

Orientation and Mobility Issues. This family of challenges is the most thoroughly discussed in the current literature, since it directly arises from the metaverse intrinsically being based on the immersive experience of mobility in the digital realm. In fact, interacting with components like virtual reality controllers often requires precise actions, such as pressing small buttons, or performing specific arm, hand, and finger gestures. This presents challenges for individuals with disabilities who may not be able to perform these actions in the same way or to the same extent as the technology expects. For example, individuals with motor limitations may face difficulties operating VR controllers, and those with spinal cord injuries may struggle to wear headsets, which are quite heavy. Furthermore, people who use crutches or wheelchairs may find it difficult or even impossible to use VR controllers available at the state of the art, since VR technologies often require full-body engagement.

Inadequate User Interface. Inadequacies in the user interface lacking proper feedback further complicate the participation in the metaverse for individuals with sensory or motor impairments, as it can hinder their understanding of available platform functionalities and operations. For instance, if a specific arm is required for mid-air pointing and selection, it can pose difficulties for individuals who prefer to use their stronger or potentially different arm. Similarly, actions that demand a combination of speed and precision can be problematic for individuals with tremors.

Avatar Customization. A further issue faced by individuals with disabilities when accessing the metaverse is related to avatar customization. Upon entering the metaverse, users have the option to choose their avatar's appearance, customizing various aspects such as facial features, hair colour, clothing, and accessories. Consequently, users with disabilities should have the opportunity to decide whether or not to represent their disability, for example, by choosing to depict their avatar in a wheelchair.

⚐ Answer to RQ₁. The main kinds of challenges faced by people with disabilities who are interested in enjoying the metaverse experience are related to lack of alternative textual descriptions and appropriate markups, orientation and mobility issues, inadequate user interface, and avatar customization. Some of the challenges are due to the intrinsic nature of the metaverse, e.g., moving in the Virtual Reality, others are caused by sub-optimal implementation choices, e.g., inadequate feedback provided by the user interface.

Although the metaverse is currently subject to massive hype coming from both industries and enthusiastic people, it is not free from issues that undermine the experience of individuals with disabilities. This raises the need for practitioners involved in the design and development of metaverse platforms to pay particular attention to accessibility and inclusivity, to avoid letting people out from this novel realm.

⊟ Takeaway Message 1. There is the need for stakeholders involved in the creation of metaverse realms to acknowledge current accessibility issues and correct them to grant all people to enjoy the experience independently from their disabilities.

4.2   RQ₂: On the Countermeasures to Allow People with Disabilities to Enjoy the Metaverse

After gathering information on the open issues and challenges experienced by people with disabilities when trying to access and interact within the metaverse, our second research question encouraged us to look for possible solutions to deal with such problems and guarantee accessibility to all individuals. A number of six papers contributed to answering such research question [7,9,11,13,16,21], reporting accessibility proposals related to three disabilities, i.e., motor, visual, and auditory impairments.

Motor Disabilities. The family of orientation and mobility issues is the one that mostly concerns people with motor disabilities. To overcome such difficulties, a number of solutions have been proposed in the literature, aiming at helping individuals interact within the metaverse realm. Some examples include (1) a pen-like controller enabling input through wrist movement, (2) gaze- or voice-based command interfaces which do not require a direct match between physical movements and virtual ones, and (3) non-invasive biosensors to detect eye movements and facial expressions, enabling alternative modes of interaction.

Visual Disabilities. People with visual impairments necessitate audio- and tactile-based interfaces to gain an enjoyable experience in the metaverse. The literature suggested the use of creative solutions to enable a proper interaction, such as (1) tactile gloves improving object perception, (2) *canetrollers*, i.e., controllers mimicking the experience of using a white cane in virtual reality, which in turn need the enhancing of markups in the digital realm to provide acoustic feedback to users, and (3) toolkits to manipulate the surrounding environment, such as magnification, brightness and contrast adjustment, and object recognition.

Auditory Disabilities. Orientation issues mainly impact people with auditory disabilities, who may experience difficulties in perceiving the direction of sound in the digital realm. To overcome such struggle, the literature proposed a system which is capable of identifying three-dimensional sounds and translating them into vibrations indicating the direction of the sound source. Vibration motors are positioned inside the user's ears, and the frequency of vibrations varies between the left and right ears to denote the origin of the sound from different directions; in this way, the difference perceived within the two ears enables users to accurately recognize the direction of the sound.

> ✍ **Answer to RQ₂.** Accessibility of the metaverse can be improved by intervening in the underlying technologies, such as virtual reality. Examples of solutions encompass alternative input methods, strengthening of markers, and tactile and acoustic feedback.

Although a number of solutions have been proposed to assist people with physical disabilities, further effort is still needed to allow individuals with other kinds of impairments to enjoy the experience in the metaverse.

> 🖴 **Takeaway Message 2.** Further solutions are needed to aid people with other kinds of cognitive disabilities, such as learning disorders, communication impairments and difficulties in social interaction.

5   Conclusion

In this paper, we performed a Systematic Literature Review to assess the state of the art on accessibility in the metaverse, focusing on the challenges and related solutions for people with disabilities. Despite the considerable progress in the technologies enabling the metaverse realm, individuals with physical and cognitive impairments may have trouble interacting within the digital world, due to lack of appropriate markups and alternative descriptions, inadequate user interface, and avatar customization choices. A number of solutions have been proposed in the literature to overcome some issues, such as alternative controllers, interaction methods, and feedback techniques. However, there is still the need for further research effort to identify assistive technologies for people with psychological and social difficulties, who may struggle interacting with peers in the metaverse realm, thus feeling frustrated or stressed also in the real world. These findings can serve as valuable input for developing AI tools to support people, including students, with disabilities in navigating the metaverse.

References

1. Lopez, G.A.M., et al.: The university in the metaverse. Proposal of application scenarios and roadmap model. In: 2022 Congreso de Tecnología, Aprendizaje y Enseñanza de la Electrónica (XV Technologies Applied to Electronics Teaching Conference), pp. 1–9. IEEE (2022)
2. Basilaia, G., Kvavadze, D.: Transition to online education in schools during a SARS-CoV-2 coronavirus (COVID-19) pandemic in Georgia. Pedagogical Res. **5**(4) (2020)
3. Bilotti, U., Di Dario, D., Palomba, F., Gravino, C., Sibilio, M.: Machine learning for educational metaverse: how far are we? In: 2023 IEEE International Conference on Consumer Electronics (ICCE), pp. 1–2. IEEE (2023)
4. Braun, V., Clarke, V.: Using thematic analysis in psychology. Qual. Res. Psychol. **3**(2), 77–101 (2006)
5. Di Dario, D., Sellitto, G., Pentangelo, V., Fede, M.L., Ferrucci, F.: Breaking barriers in the metaverse: a comprehensive exploration of accessibility for users with disabilities – online appendix (2023). https://doi.org/10.6084/m9.figshare.24549106
6. Duan, H., Li, J., Fan, S., Lin, Z., Wu, X., Cai, W.: Metaverse for social good: a university campus prototype. In: Proceedings of the 29th ACM International Conference on Multimedia, pp. 153–161 (2021)
7. Dudley, J., Yin, L., Garaj, V., Kristensson, P.O.: Inclusive immersion: a review of efforts to improve accessibility in virtual reality, augmented reality and the metaverse. Virtual Reality **27**(4), 2989–3020 (2023)
8. Gerling, K., Spiel, K.: A critical examination of virtual reality technology in the context of the minority body. In: Proceedings of the 2021 CHI Conference on Human Factors in Computing Systems, CHI 2021. Association for Computing Machinery, New York (2021)
9. Hadi Mogavi, R., Hoffman, J., Deng, C., Du, Y., Haq, E.U., Hui, P.: Envisioning an inclusive metaverse: student perspectives on accessible and empowering metaverse-enabled learning. In: Proceedings of the Tenth ACM Conference on Learning @ Scale. L@S 2023. Association for Computing Machinery, New York (2023)

10. Jeanne Choi, H.W.K.: Empirical evaluation of metaverse accessibility for people who use alternative input/output methods (2023)
11. Kanza, Y., Krishnamurthy, B., Srivastava, D.: Geospatial accessibility and inclusion by combining contextual filters, the metaverse and ambient intelligence. In: Proceedings of the 30th International Conference on Advances in Geographic Information Systems. SIGSPATIAL 2022. Association for Computing Machinery, New York (2022)
12. Kitchenham, B.: Procedures for performing systematic reviews. Keele, UK, Keele University, vol. 33, no. 2004, pp. 1–26 (2004)
13. Meena, S.D., Mithesh, G.S.S., Panyam, R., Chowdary, M.S., Sadhu, V.S., Sheela, J.: Advancing education through metaverse: components, applications, challenges, case studies and open issues. In: 2023 International Conference on Sustainable Computing and Smart Systems (ICSCSS), pp. 880–889 (2023)
14. Mystakidis, S.: Metaverse. Encyclopedia **2**(1), 486–497 (2022)
15. Mystakidis, S.: Metaverse. Encyclopedia **2**(1), 486–497 (2022). https://doi.org/10.3390/encyclopedia2010031. https://www.mdpi.com/2673-8392/2/1/31
16. Parker, C., et al.: Towards an inclusive and accessible metaverse. CHI EA 2023. Association for Computing Machinery, New York (2023)
17. Pellas, N., Dengel, A., Christopoulos, A.: A scoping review of immersive virtual reality in stem education. IEEE Trans. Learn. Technol. **13**(4), 748–761 (2020)
18. Quinlan, M.: Disconnected from reality: do the core concepts of the metaverse exclude disabled individuals? arXiv preprint arXiv:2303.08222 (2023)
19. Radanliev, P., De Roure, D., Novitzky, P., Sluganovic, I.: Accessibility and inclusiveness of new information and communication technologies for disabled users and content creators in the metaverse. Disabil. Rehabil. Assist. Technol. 1–15 (2023)
20. Ritterbusch, G.D., Teichmann, M.R.: Defining the metaverse: a systematic literature review. IEEE Access **11**, 12368–12377 (2023)
21. Seigneur, J.M., Choukou, M.A.: How should metaverse augment humans with disabilities? In: 13th Augmented Human International Conference. AH 2022. Association for Computing Machinery, New York (2022)
22. Stephenson, N.: Snow Crash. Bantam Books, New York (1992)
23. Whittaker, M., et al.: Disability, bias, and AI. AI Now Institute, vol. 8 (2019)
24. Zhang, K., Deldari, E., Lu, Z., Yao, Y., Zhao, Y.: "it's just part of me:" understanding avatar diversity and self-presentation of people with disabilities in social virtual reality. In: Proceedings of the 24th International ACM SIGACCESS Conference on Computers and Accessibility, pp. 1–16 (2022)

The Impact of Music and Metaverse on Education: Results of a Scoping Review

Alessio Di Paolo$^{(\boxtimes)}$ (iD), Michele Domenico Todino(iD), and Maurizio Sibilio(iD)

Department of Humanities, Philosophy and Education, University of Salerno, Fisciano, Italy
{adipaolo,mtodino,msibilio}@unisa.it

Abstract. The idea of the metaverse, a virtual reality-based digital universe, has garnered significant attention due to its potential to revolutionize education and music-related activities. This paper explores the multifaceted relationship between the metaverse, education, and music, shedding light on the opportunities and challenges it presents. Using the PRISMA-ScR method, we proceeded to a sampling of the contributions currently present on the net and extrapolated from search engines such as Eric, Scopus, JSTOR, Google Trends, ACM, Sage Journal, and WorldCat. Based on previously selected indicators, 175 scientific contributions were analyzed, among which those most relevant to the topic were selected based on previously chosen indicators. A clear interest in the relationship between metaverse and music emerges from the contributions, without, however, focusing much attention on the impact these realities can have on the educational process.

Keywords: Music · Metaverse · Education

1 Introduction

The theme of the metaverse and its possible application for education involved several studies [1, 2]. In particular, the aim is to verify in what terms it can contribute to strengthening the social, and communicative sphere, intervening to speed up dialogue and make it more effective for different cognitive styles, contributing to an improvement in educational processes, especially for the possibility to immersive experiences that it offers [3, 4]. Music, in this *vicariant sense* [5, 6] sense, has proven to be an effective means for motor, perceptual, cognitive [7], and socio-communicative development functional to the path of personal growth, as well as useful for achieving autonomy and self-awareness [8]. It is, therefore, interesting to verify what interconnection there may be between music and the metaverse, to increase learning processes.

The scoping review conducted had precisely this objective. The text will present the results with the first part of the paper dedicated to a reflection on the metaverse and its

---

The article is the result of the scientific comparison and collaboration of the authors. However, the assignment of scientific responsibility is as follows: Alessio Di Paolo is the author of paragraphs 2. "The relationship between metaverse and education," 3. "Metaverse and music," and 4. "Methodology"; Michele Domenico Todino is co-author of paragraphs 1. "Introduction" and 5. "Conclusions." Maurizio Sibilio is scientific coordinator of the project.

© The Author(s), under exclusive license to Springer Nature Switzerland AG 2024
F. Palomba and C. Gravino (Eds.): WAILS 2024, LNCS 14545, pp. 56–66, 2024.
https://doi.org/10.1007/978-3-031-57402-3_7

potential for educational processes; a second part of the paper, instead, will be destined to analyze the relationship between music and metaverse; the last part of the paper will present the results of the scoping review. It should be specified that the metaverse being researched is understood in a broad sense, that is, as shared three-dimensional virtual spaces inhabited by people through digital avatars, in which it is possible to interact with the environment and other users in real time, not focusing on the implementation proposed in Zuckerberg's project [9].

2   The Relation Between Metaverse and Education

The concept of the metaverse, a virtual reality-based digital universe where individuals interact with each other and digital environments, has been a topic of great interest and speculation in recent years. As technology continues to advance, the metaverse is becoming more tangible, and its potential applications are numerous. One of the most promising and transformative areas where the metaverse can make a significant impact is education. [10] In this comprehensive exploration of the relationship between the metaverse and education, we will delve into various aspects of this connection, including the potential benefits, challenges, and implications for the future of learning [11].

The metaverse is a term often used to describe a collective virtual shared space, merging physical and digital reality [4], populated by avatars representing users from all around the world [12]. It goes beyond augmented and virtual reality by creating a persistent and interconnected environment that can be accessed via various devices, not just headsets. In the metaverse, users can communicate, socialize, work, play, and learn as if they were physically present. As a result, the metaverse presents a unique opportunity to revolutionize education [13].

The metaverse has the potential to *democratize education* by making it accessible to people worldwide [14], regardless of their physical location or socioeconomic status. Traditional education often requires students to be physically present in a specific location, but the metaverse erases these geographical barriers [15]. The root of the term μετα-" ("through, beyond, after") implies a process through which the user becomes able to overcome the limits and barriers posed by the universe itself, opening new cognitive and experiential horizons, including those offered by virtual.

All that is needed is an internet connection, and students from different corners of the world can come together in a shared virtual classroom.

In the metaverse, educational experiences can be highly personalized. AI-driven algorithms can assess a student's strengths, weaknesses, and learning style, tailoring the curriculum *to connect* individual needs [16]. This *adaptive approach* can significantly enhance learning outcomes and help students progress at their own pace. To fully harness, furthermore, its benefits while mitigating associated challenges, several key considerations need to be addressed.

Firstly, establishing ethical and legal frameworks to protect students and their data within the metaverse is paramount [16]. This includes crafting regulations for virtual classrooms and intellectual property rights. As education increasingly relies on digital environments, safeguarding the privacy and rights of students is crucial. Moreover, integration with existing educational systems is essential for a smooth transition and consistent learning outcomes [2].

The metaverse should seamlessly connect with current curricula and tools to ensure that the new technology enhances, rather than disrupts, the educational process. Ongoing research and assessment of the metaverse's impact on education are necessary. This entails a commitment to understanding its efficacy and making improvements as needed. Monitoring and adapting to the changing landscape of educational technology will be key to optimizing the metaverse's potential. The metaverse can revolutionize various aspects of education [10]. An example could be *virtual classrooms*. Teachers can conduct classes in immersive virtual environments, providing students with interactive and engaging learning experiences. A second modality of work using immersive experiences in the educational field could be *laboratory Simulations. S*cience and engineering students can conduct experiments and simulations in virtual labs, reducing the cost and risk associated with physical experiments. The mediation of virtuality is useful also for the physical safety of workers.

The use of virtual experiences is important also for *historical reenactments*. History lessons can come alive through time-travel experiences, allowing students to witness and immerse themselves in historical events. The use of immersive reality is also a valid support for improving *language skills.* Virtual language immersion can help students practice speaking and listening in a foreign language with native speakers, enhancing language acquisition. It is important to consider, furthermore, the inclusive potential of immersive experiences. The metaverse can offer customized learning experiences for students with special needs, making education more accessible, adaptable, and engaging.

Several educational institutions and organizations are already exploring the metaverse's potential [17]. We might think, for example, of *Instructure*, which is exploring ways to enhance online education using the metaverse, building on its experience with the learning management system Canvas, or *Mursion,* which offers virtual reality simulations for teacher training, allowing educators to hone their teaching skills in realistic virtual classrooms. *Engage,* moreover provides a range of educational VR experiences, from historical reenactments to virtual field trips and language learning opportunities. Lastly, *AltspaceVR*, is a social VR platform where educators can host events, conferences, and classes, facilitating interaction and engagement in a virtual space.

The metaverse offers immersive, personalized, and globally accessible learning experiences that can empower students of all backgrounds [18]. However, it also presents challenges that need to be addressed, such as privacy concerns, digital literacy, and the quality of educational content. Therefore, a thoughtful, ethical, and well-regulated approach is necessary to harness the metaverse's benefits while ensuring the well-being and progress of learners in this digital age.

As we move forward, educational institutions, technology companies, and policy-makers need to work together to harness the full potential of the metaverse in education. By doing so, we can create a future where learning knows no geographical boundaries and every student has access to a world of knowledge and experiences within the metaverse.

3 Metaverse and Music

Throughout history, the landscape of music composition, performance, education, and consumption has undergone continuous transformation, shaped by technological advancements and the ever-changing tastes of musicians and audiences. Today, a novel frontier known as the metaverse is emerging, offering a new realm for musical activities. The metaverse envisions a digital, virtual world that runs parallel to our physical reality, where users navigate and interact through personalized avatars [19]. This vision builds upon extensive research in academia and industry, exploring immersive technologies, gaming platforms, and virtual spaces for social interaction [20].

Defining the metaverse in the literature remains a point of contention. The metaverse transcends the limitations of time and space, offering users an immersive experience. It represents a significant evolution of the internet, enabling connected users to interact virtually, and fostering the sensation of sharing the same environment. This digital realm has the potential to facilitate work, commerce, collaboration, socialization, creativity, learning, and recreation. However, it's important to note that this vision is still in its infancy, despite the enthusiastic support from major tech companies promoting their visions of future business lines.

Presently, various types of metaverses exist, such as gaming-based ones (e.g., Fortnite, Roblox, Second Life, or Minecraft) and blockchain-based metaverses (e.g., Decentraland or The Sandbox) [10]. Some metaverses, like Horizon by Meta, are built entirely around VR technologies, necessitating VR headsets for full immersion. All these metaverse types share a common thread: *they provide a virtual space for real-time interactions with the digital environment and other users in the form of* avatars.

The metaverse presents the opportunity to foster social connections and engage in entertainment. Music, as one of the most popular forms of entertainment, has seen the establishment of numerous virtual clubs and concert halls in recent years to cater to this demand [21]. These digital spaces allow people to congregate, form friendships, dance, and enjoy live music or recorded tunes. A wide array of musical activities can be carried out in the metaverse, spanning from composition and performance to recreational music creation, teaching, and experiencing virtual live concerts. These developments have significant implications both at an artistic and commercial level.

While the field is rapidly evolving, the "Music Metaverse" (MM) concept remains in its early stages of development. It is a constantly evolving idea, with different musical stakeholders contributing to its definition in various ways. However, to the best of the author's knowledge, no comprehensive investigation into the opportunities and challenges of using the metaverse for musical activities has been conducted thus far.

In the past decade, authors have discussed the musical aspects of the metaverse, particularly focusing on platforms like Second Life [22, 23]. However, during that time, the technology for XR and networked interactions was in its infancy, social networks were only beginning to gain mass adoption, digital currencies were not widespread, and systems for music broadcasting, streaming, and networked performances were suboptimal. Today, the technological and non-technological landscape is radically different, with the COVID-19 pandemic accelerating the demand for online musical interaction [24].

Over the past two decades, there has been a rapid expansion in the field where Extended Reality (XR) and music converge, giving rise to a well-established area of research known as "Musical XR" [25]. Various authors have explored different facets of this broad domain [26, 27].

The integration of XR technologies into musical activities marks a paradigm shift, disrupting traditional modes of musical interaction by enabling performers and audiences to engage musically with virtual/augmented objects, agents, and environments. Research has also delved into the perception aspects of Musical XR settings. Another field of application of music and metaverse is related to the Internet of Musical Things. The concept of the Internet of Musical Things (IoMusT) within the broader Internet of Things (IoT) landscape specifically targets the musical domain. It encompasses a network of Musical Things designed for musical purposes, including smart musical instruments and wearable devices for performers and audiences [28].

Real-time musical interactions in networked Musical XR environments remain relatively uncharted territory, with only a few studies exploring collaborative music-making in XR and even fewer focusing on interactions mediated by the network [21]. Interesting is, also Blockchain technology. Blockchain technology, on the other hand, revolves around a decentralized system of interconnected data units, known as blocks. It serves as a distributed ledger shared among independent nodes of a computer network, recording transactions chronologically. Notably, blockchain *enables peer-to-peer transactions* without relying on centralized authorities or third parties, ensuring the confidentiality and integrity of exchanged data through cryptographic mechanisms. Blockchain technology offers various applications in the music industry, including creating a networked database of copyright ownership to address issues with music licensing. It can also facilitate the management of royalties through smart contracts, leading to immediate and transparent royalty payments to artists. This technology can enhance transparency within the value chain by addressing issues related to the processing of payments by record labels, management organizations, or publishers, ensuring that artists receive their rightful royalties. However, these use cases primarily concern challenges in the recording music industry, which represents just one segment of the music sector [29].

4   Methodology

This contribution aims to investigate the current relationship between music, metaverse, and education, to understand whether there are scientific studies that have already recognized how music, applied to metaverse, can contribute to the growth educational process. To achieve this goal, the authors decided to make a preliminary systematic review by adopting the PICO conceptual framework [30]. Moreover, based on suggestions from Hume's research team [31], this review followed four phases: research, screening, data mining, and comprehensive review of the included studies [32]. The PICO conceptual framework will be used as it invites you to pay attention to specific elements that are summarized through the acronym: P for *Patient or problem* that is main characteristics of the patient; problem or condition coexisting with a particular pathology; I for *Intervention* that is type of intervention/practice adopted; C for *Comparison* that is alternative interventions; R for Result that is expected results or effects.

During the first phase, a literature search was conducted using eight international databases (ACM Digital Library, Educational Resource Information Center (ERIC), Google Scholar, Google Trends, JStor, Sage Journal, Scopus, and Worldcat). The articles considered were those published (or pre-published online) in a peer-reviewed journal until 01/01/2023.

To narrow the search and retrieve all bibliographic resources, referring exclusively to the macro-categories, it was decided to use only the Boolean operator "AND". The keywords identified for the search and connected with the Boolean connector "AND" were "Music, Metaverse, Education" In addition, the systematic review process included the search for already published, peer-reviewed, and full-text studies in English.

During the second phase, all duplicates were removed, and the most relevant contributions were identified based on the analysis of the title and, subsequently, the abstract.

Following this selection, the contributions were downloaded and further screened according to the inclusion and exclusion criteria (Table 1) defined according to the suggestions of the PICO conceptual framework and the method used by Hume et al.

Table 1. Inclusion and exclusion criteria for the analysis of contributions

Category	Inclusion criteria	Exclusion criteria
Literature	Article published (or pre-published online) in a *peer-reviewed journal*	Grey literature (thesis or doctoral thesis, proceedings, …)
Tongue	English	Other languages
Drawing	Experimental group study and individual cases	Systematic reviews, meta-analysis, scope review
Population/Participants (P)	Children from 0 to 15 years	Older participants
Intervention (I)	Evidence-based practice involves the interconnected use of music and metaverse for inclusion	Only Medical or psychopharmacological interventions
Comparison (C)	Experimental platforms on the metaverse or specific tools	Other type of platforms
Results (O)	Results on improving the educational process	Only the results of parents or caregivers of other skills

All studies that met the inclusion criteria were independently analyzed by the authors and critically discussed in the conclusion of the analysis. The summary and analysis of the articles have been summarized in a table structured as follows: (I) author/year of publication/title, (II) sample, (III) research objectives; and (V) improvements or other significant results.

The bibliographic research initially produced 175 results of contributions, distributed as follows. ACM Digital Library returned 15 contributions; Educational Resource Information Center (ERIC) returned 1 result; Google Scholar returned 115 results; Google trend returned 0 results; Jstor returned 26 results; Sage Journal returned 3 results; Scopus returned 10 results; Worldcat returned 5 results.

Following this finding, during the screening phase, duplicates were eliminated (no. 6) and contributions were selected based on their relevance about the title (no. 9 = relevant; no. 160 = not relevant) and, consequently, to the abstracts (no. 5 = relevant; no. 4 = not relevant).

During the third phase, the data of each contribution were extracted, considering the inclusion and exclusion criteria previously explained. The extraction revealed that, among the studies analyzed, those relevant and eligible for analysis were n. 3, and those not relevant 2 (Fig. 1).

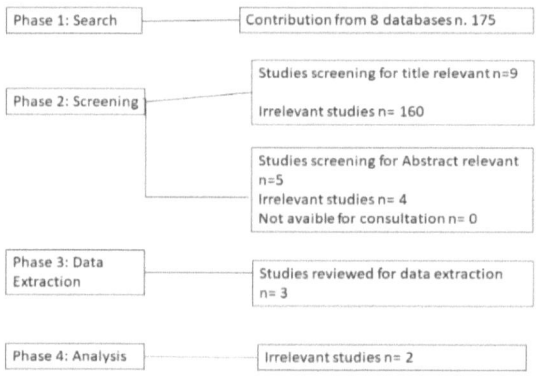

Fig. 1. PRISMA flow chart

5   Results

Regarding the results emerged that the three contributions discuss various aspects of education, technology, and culture, with a particular focus on music therapy, metaverse-based learning, and music appreciation within the context of advanced computer technology and virtual spaces. Let's relate these abstracts and their key points. Lin & Liao [33] in this sense, emphasize the promotion and innovation of music and education in the metaverse. Their project emphasizes the use of professionally established AI music classrooms equipped with high-quality technology and excellent teachers. This approach aims to meet international standards and encourages the popularity of classical instruments, music theory, music therapy, and AI courses. The authors also highlight how an AI music economy industry is being generated that integrates Western, international, and metaverse cultures with cultural elements. The idea is to not retard the development of national and Aboriginal music and art, ultimately promoting a sense of beauty and creativity in the post-epidemic era.

Dreamson & Park [34] point out how the metaverse can be related to other technological terms such as augmented reality, lifelogging, mirror worlds, and virtual worlds, and the use of a metaverse platform is considered an emerging form of learning. They also analyze how pedagogical features, such as self-learning, collaborative learning, and learning-by-doing, are similar in virtual and real learning. Metaverse-based learning refers to a new reality that learners create by assuming that the existence of reality is different in the metaverse. Music becomes a design artifact, a setting for activities and reflection and presentation that fosters collaborative learning with co-ownership and co-authorship; interconnectedness with all living and non-living beings; co-participants with different roles; and research-driven transdisciplinary learning, which is used to articulate a structure for metaverse-based learning that differs in its 'community' character.

Interesting the study of Lee & Hwang [35], which propose the use of the metaverse platform SPOT for music appreciation lessons. It examines the characteristics of the metaverse, the features of SPOT, and its utilization for music appreciation. The study proposes a teaching and learning model for music appreciation within the metaverse, creating a virtual space for students to engage in diverse types of music appreciation without constraints of time and space. This approach is shown to enhance student engagement and learning outcomes, transcending traditional boundaries of music reception and expression, learning space and time, and real and virtual activities.

Overall, these contributions present a common thread of harnessing advanced technology, particularly metaverse platforms, in the fields of education and culture, whether for music therapy, metaverse-based learning, or music appreciation. These innovations seek to enhance the learning experience, promote creativity, and link cultural elements with cutting-edge technologies.

The three contributions also provide valuable insights into the intersection of education, technology, and culture, particularly in the context of the metaverse. However, although the topic of the metaverse has been explored extensively in recent years, they do not present a precise focus on the relevance that music could have on education transposed to this alternative but equally effective educational universe, which could instead be considered for innovative didactics, more for everyone and everyone's benefit, from an inclusive perspective.

6   Conclusions

The metaverse presents significant potential for revolutionizing education through immersive learning experiences, democratizing access to education, and personalizing learning for individual students, especially with SNE [36, 37]. However, challenges such as technical requirements, privacy, addiction, digital literacy, and content quality need to be addressed. Educational institutions must invest in technology, provide teacher training, develop high-quality content, establish ethical and legal frameworks, and seamlessly integrate the metaverse into existing systems. Ongoing research and assessment are essential to understand the metaverse's impact on education and make necessary improvements. The metaverse is expected to lead to virtual classrooms, laboratory simulations, historical reenactments, language learning, career training, and special education. Several educational institutions and organizations are already exploring metaverse-based learning [38, 39].

The metaverse introduces a new frontier for music activities, creating a "Music Metaverse" (MM) dedicated to music-related experiences. Music, as a popular form of entertainment, has seen the creation of virtual clubs and concert halls in the metaverse, allowing people to gather, socialize and enjoy live music or recorded tracks. The MM offers opportunities for musical composition, performance, education, and social interaction, with important artistic and commercial implications. Various types of metaverse, such as game-based and blockchain, support music-related activities. The MM concept is in the early stages of development and needs further exploration. The integration of XR technologies into musical activities, also known as 'Musical XR', enables new musical interactions and experiences. Blockchain technology and non-fungible tokens (NFT) are transforming the music industry and offering musicians new opportunities for audience engagement. Digital tokens can also play a key role in MM by creating virtual representations of musical entities and environments. In conclusion, it would be desirable to continue investigating the potential of music in the metaverse for education, emphasizing the potential benefits, challenges, and opportunities arising from the integration of advanced technologies in these areas.

Disclosure of Interests. The authors have no competing interests to declare that are relevant to the content of this article.

References

1. Inceoglu, M.M., Ciloglugil, B.: Use of Metaverse in education. In: Gervasi, O., Murgante, B., Misra, S., Ana, M.A., Rocha, C., Garau, C. (eds.) Computational Science and Its Applications – ICCSA 2022 Workshops: Malaga, Spain, July 4–7, 2022, Proceedings, Part I, pp. 171–184. Springer, Cham (2022). https://doi.org/10.1007/978-3-031-10536-4_12
2. Hwang, G.J., Chien, S.Y.: Definition, roles, and potential research issues of the metaverse in education: an artificial intelligence perspective. Comput. Educ. Artif. Intell. **3**, 100082 (2022)
3. Aiello, P., Pace, E.M., Sibilio, M.: A simplex approach in Italian teacher education programs to promote inclusive practices. Int. J. Inclusive Educ. **27**(10), 1163–1176 (2023)
4. Sibilio, M., et al.: MetaWelt: embodied in which body? Simplex didactics to live the web 3.0. In: Antona, M., Stephanidis, C. (eds.) Universal Access in Human-Computer Interaction, HCII 2023. LNCS, vol. 14021, pp. 111–119. Springer, Cham (2023). https://doi.org/10.1007/978-3-031-35897-5_8
5. Berthoz, A.: The Vicarious Brain, Creator of Worlds. Harvard University Press, Harvard (2017)
6. Sibilio, M.: La semplessità: proprietà e principi per agire il cambiamento, Morcelliana, Brescia (2023)
7. Proverbio, A.M.: Neuroscienze cognitive della musica. Il cervello musicale tra arte e scienza, Zanichelli, Milano (2019)
8. Draper, P.: Music two-point-zero: music, technology and digital independence. J. Music Technol. Educ. **1**(2–3), 137–152 (2008)
9. We believe in the future of connection in the Metaverse. https://about.meta.com/metaverse/. Accessed 11 Nov 2023
10. Jin, C., Wu, F., Wang, J., Liu, Y., Guan, Z., Han, Z.: MetaMGC: a music generation framework for concerts in Metaverse. EURASIP J. Audio Speech Music Process. **2022**(1), 31 (2022)

11. Zhang, X., Chen, Y., Hu, L., Wang, Y.: The Metaverse in education: definition, framework, features, potential applications, challenges, and future research topics. Front. Psychol. **13**, 6063 (2022)
12. Mystakidis, S.: Metaverse. Encyclopedia **2**(1), 486–497 (2022)
13. Azoury, N., Hajj, C.: Perspective chapter: the Metaverse for education (2023)
14. Pickup, A., Heybach, J.: Social media's education grab. Philanthrocapitalism, data centres and the Metaverse vision of education. In: World Yearbook of Education 2024: Digitalisation of Education in the Era of Algorithms, Automation and Artificial Intelligence, p. 102 (2023)
15. Hudson-Smith, A., Batty, M.: Ubiquitous geographic information in the emergent Metaverse. Trans. GIS **26**(3), 1147–1157 (2022)
16. Mourtzis, D., Panopoulos, N., Angelopoulos, J., Wang, B., Wang, L.: Human centric platforms for personalized value creation in Metaverse. J. Manuf. Syst. **65**, 653–659 (2022)
17. Tayal, S., Rajagopal, K., Mahajan, V.: Virtual reality based Metaverse of gamification. In: 2022 6th International Conference on Computing Methodologies and Communication (ICCMC), pp. 1597–1604. IEEE, Spain (2022)
18. Stanoevska-Slabeva, K.: Opportunities and challenges of Metaverse for education: a literature review. In: EDULEARN22 Proceedings, pp. 10401–10410 (2022)
19. Danowski, P.: Sound of the Metaverse: a breif history of virtual reality music instruments and virtual music venues (2021)
20. Lee, L.H., et al.: When creators meet the Metaverse: a survey on computational arts. arXiv preprint arXiv:2111.13486 (2021)
21. Turchet, L.: Musical Metaverse vision, opportunities, and challenges. Pers. Ubiquit. Comput. **27**, 1811–1827 (2023). https://doi.org/10.1007/s00779-023-01708-1
22. Armitage, J.: Rethinking haute couture: Julien Fournié in the virtual worlds of the Metaverse. Fr. Cult. Stud. **34**(2), 129–146 (2023)
23. Kumar, A.: Immersive 3D Design Visualization: With Autodesk Maya and Unreal Engine 4. Apress, Berkeley (2021)
24. Johnson, D.D.: Generating polyphonic music using tied parallel networks. In: Correia, J., Ciesielski, V., Liapis, A. (eds.) Computational Intelligence in Music, Sound, Art and Design: 6th International Conference, EvoMUSART 2017, Amsterdam, The Netherlands, April 19–21, 2017, Proceedings, pp. 128–143. Springer, Cham (2017). https://doi.org/10.1007/978-3-319-55750-2_9
25. Bowman, S.R., Vilnis, L., Vinyals, O., Dai, A.M., Jozefowicz, R., Bengio, S.: Generating sentences from a continuous space. arXiv preprint arXiv:1511.06349 (2015)
26. Serafin, S., Erkut, C., Kojs, J., Nilsson, N.C., Nordahl, R.: Virtual reality musical instruments: state of the art, design principles, and future directions. Comput. Music. J. **40**(3), 22–40 (2016)
27. Berthaut, F., Zappi, V., Mazzanti, D.: Scenography of immersive virtual musical instruments. In: 2014 IEEE VR Workshop: Sonic Interaction in Virtual Environments (SIVE), pp. 19–24. IEEE, Spain (2014)
28. Lin, H., Wan, S., Gan, W., Chen, J., Chao, H.C.: Metaverse in education: vision, opportunities, and challenges. In: 2022 IEEE International Conference on Big Data (Big Data), pp. 2857–2866. IEEE (2022)
29. Arcos, L.C.: The blockchain technology on the music industry. Braz. J. Oper. Prod. Manage. **15**(3), 439–443 (2018)
30. Richardson, W.S., Wilson, M.C., Nishikawa, J., Hayward, R.S.: The well-built clinical question: a key to evidence-based decisions. ACP J. Club **123**(3), A12–A13 (1995)
31. Hume, K.: Evidence-based practices for children, youth, and young adults with autism: third generation review. J. Autism Dev. Disord. **51**, 4013–4032 (2021)
32. Davies, K.S.: Formulating the evidence based practice question: a review of the frameworks. Evid Based Libr Inf Pract **6**, 75–80 (2011)

33. Lin, S.Y., Liao, C.H.: The promotion and innovation of modern music education culture. Int. J. Innov. Appl. Soc. Sci. Eng. Technol. **3**(3), 19 (2022)
34. Dreamson, N., Park, G.: Metaverse-based learning through children's school space design. Int. J. Art Des. Educ. **42**(1), 125–138 (2023)
35. Lee, H., Hwang, Y.: Technology-enhanced education through VR-making and Metaverse-linking to foster teacher readiness and sustainable learning. Sustainability **14**(8), 4786 (2022)
36. Di Tore, S., Campitiello, L., Todino, M. D., Iannaccone, A., Sibilio, M.: Education in the Metaverse: amidst the virtual and reality. Ital. J. Health Educ. Sport Incl. Didactics **6**(3) (2022)
37. Di Tore, P.A., Axelsson, E.E.P.: Metaverse, between ecology of virtual and phenomenology of experience. J. Inclusive Methodol. Technol. Learn. Teach. **3**(1) (2023)
38. Di Tore, S.: Dal metaverso alla stampa 3D: Prospettive semplesse della didattica innovativa, Studium edizioni, Roma (2022)
39. Todino, M.D., Campitiello, L., Di Tore, S.: Simplexity to improve human-machine interaction in 3D virtual reality. Int. J. Digit. Lit. Digit. Competence (IJDLDC) **14**(1), 1–8 (2023)

Future Support Teachers' Perceptions and Opinions on Effective Learning Through the Promotion of ICT

Ilaria Viola⬥, Emanuela Zappalà(✉)⬥, and Maurizio Sibilio⬥

University of Salerno, Salerno, Italy
{iviola,ezappala}@unisa.it

Abstract. The adoption of Artificial Intelligence (AI) within the field of Information and Communication Technologies (ICT) has transformative potential in the field of education, particularly in special education. In fact, ICT is still employed to enhance and personalize teaching practices to meet students' educational needs. Considering that several investigations showed that technological skills and the usage of artificial intelligence are inextricably linked, as technological skills are often essential for fully implementing the potential of AI, the contribution aims to explore perceptions and opinions of future support teachers on the use of technology in education trough administration of a modified version of the SELFIE questionnaire.

Keywords: Artificial Intelligence · ICT · teachers' perceptions · teachers' opinions · inclusive processes

1 Introduction

The use of technology in classroom has become increasingly prevalent in recent years, especially with the integration of Information and Communication Technology (ICT) and Artificial Intelligence (AI) in education. Both ICT and AI are playing a central role in education, changing the teaching interaction, altering how teachers teach and how students learn (Timotheou et al. 2023). The European Commission (2020) makes it clear, in a society of progress and fear of change, that the integration of AI in ICT aims to make education more effective and personalized, able to meet individual needs of students and to develop a set of skills that are crucial in today's world of work. In addition, AI support the personalization of education by analyzing the performance and learning needs of students to adapt teaching content and approaches (Zhang et al. 2020).

Moreover, the European Commission and other educational and research organizations (UNESCO 2023; EC 2020; UN 2015) show how the education system is progressively moving towards integrating AI into ICT to prepare students to become active

---

The article is the result of a collaboration between the authors. However, Ilaria Viola wrote paragraphs 1 "Introduction" and 3 "Conclusions"; Emanuela Zappalà wrote paragraphs 2 "Methodology", Maurizio Sibilio is the scientific coordinator of the research.

F. Palomba and C. Gravino (Eds.): WAILS 2024, LNCS 14545, pp. 67–78, 2024.
https://doi.org/10.1007/978-3-031-57402-3_8

and competent participants in the global digital society. Digital technologies are significantly reshaping education by improving pedagogical methods and fostering digital literacy among students and educators (Rogers 2014; Skryabin Zhang, Liu, & Zhang 2015). This shift entails enhancing interactive learning and teaching while equipping learners and teachers with essential digital skills. Therefore, worldwide investment by governments, schools, and families is surging to embed these technologies at the heart of educational processes, aiming to harmonize traditional education with the digital era's demands (Conrads et al. 2019). The integration of digital technologies in education has led to mixed outcomes, with investments in the field not consistently yielding substantial gains in educational productivity or cost reductions, as seen in other industries (Lim et al. 2013).

One of the most important factors on the effectiveness of the use of ICT is how teachers use these devices. For this reason, "Self-reflection on Effective Learning by Fostering Innovation through Educational technology" questionnaire was created as a self- assessment tool that guides schools through a systematic reflection process on their current use of digital technologies, acting as a catalyst to identify areas for improvement and to develop a mature approach towards digitization. Through the participation of students, teachers and school staff, the SELFIE embraces an inclusive vision that ensures a holistic understanding of the needs and perspectives of all actors involved in the educational environment. It is a concept that combines different areas of pedagogy and technology. The tool represents a significant break from traditional models of lifelong learning, promoting teacher reflection, independence, adaptability, and the development of local solutions for local problems (European Commission 2018). In the national and international landscape, SELFIE is referred to as a tool to stimulate critical reflection on teaching methods and technology adoption. Research emphasizes that SELFIE-driven self-assessment leads teachers to a greater awareness of their own digital and teaching competences (Redecker 2017), and this is crucial in addressing the challenges of contemporary education (European Commission 2018). The analysis and comparison that SELFIE stimulates, especially when integrated into in-service training programmes, allow teachers to identify specific areas where technology can enrich learning, leading to a more strategic and targeted use of digital resources (Bocconi, Kampylis, & Punie 2012).

Internationally, the framework provided by SELFIE is in tune with the principles of teacher education that emphasize the importance of innovation and continuous updating (OECD 2019).

In conclusion, the use of SELFIE in teacher education may represent an opportunity to renew the approach to teaching, integrating technologies in a way that truly serves pedagogy. The literature stresses the importance of a reflexive and critical use of ICT and AI, so that it translates into a real lever for teaching innovation and a substantial improvement of learning processes (European Commission 2018, 2023; OECD 2019). For this reason, the contribution aims to explore the opinions of future support teachers with respect to their perception of the use of technology in education.

2   Methodology

2.1   Purpose of the Study

Considering that several investigations showed that technological skills and the adoption of artificial intelligence are inextricably linked, that technological skills are often essential for fully implementing the potential of artificial intelligence (Liu et al. 2020; Sánchez-Prieto et al. 2017; Wang, Liu, Tu 2021), and that digital skills may be considered as enabling factors for the adoption of AI in classroom, the objective of this paper is to explore future support teachers' perceptions and opinions on their digital skills and on the use of technology.

2.2   Instrument

The SELFIE questionnaire, developed by the European Commission (2018), serves as a self-assessment tool for teachers and educator to evaluate their digital readiness and pedagogical practices in the context of educational technology. It encompasses various dimensions, such as digital skills, instructional methods, and the school's digital culture. Teachers may employ it for self-reflection, while schools may assess their overall digital preparedness. The questionnaire is customizable, anonymous, and confidential. Collected data may inform professional development and the integration of ICT in classroom. This initiative aligns with the European Union's efforts to enhance digital skills in educational settings.

In this study, a modified version of the SELFIE questionnaire is used to investigate the utilization of ICT in promoting inclusion in classrooms. Administered via a Google-Form, it comprises two sections:

- the first section collects socio-demographic information, including age, educational qualifications, professional position, and prior teaching experience (both in mainstream and support roles).
- the second section addresses three distinct areas corresponding to the SELFIE questionnaire:

    a) Area E and F: Pedagogy – tools, resources and classroom implementation;
    b) Area G: Assessment methods.

These sections aim to self-assess perceived digital competencies associated with technology use in the classroom. Each part consists of concise statements, rated on a 1- 5 Likert scale, and incorporates open-ended questions. Only the findings of sixteen statements and three open-ended questions will be reported in this study.

2.3   Participants and Procedure

The study employed a convenience sample comprising 123 prospective upper secondary support teachers who were enrolled in the Specialization course for educational support activities for students with disabilities at the University of Salerno (IT) during the academic year 2021/2022. Data collection took place prior to a training program leading to the acquisition of certification to work as learning support teachers in upper secondary schools. This training program consisted of multiple modules totaling 750 h, encompassing lectures, tutorials, workshops, on-site teaching practice, and tutorials. Specifically, the questionnaire was administered at the outset of the "Information and Communication Technologies" module, which spanned 75 h. All participants in the course (N = 123) were invited to complete the questionnaire voluntarily and anonymously. Among the 123 respondents, 100 (81.3%) were female, while 23 (18.7%) were male. Furthermore, the ages of the respondents ranged from under 25 to over 60, with the majority falling into the age groups of 30 to 39 (n = 43.1%) and 40 to 49 (n = 35%). Additionally, 14 (11.4%) respondents were between the ages of 50 and 59, while 11 (8.9%) were between 25 and 29 years old. Only 2 (1.6%) of the respondents were under 25 years old.

Moreover, all the respondents had varying degrees of teaching experience, with 57.8% having prior teaching experience. Specifically, 35.8% had worked as curriculum teachers, and 22% had experience as support teachers.

2.4 Data Analysis and Coding of Open-Ended Responses

Both quantitative and qualitative data were analyzed using MAXQDA® software (VERBI Software 2019). The quantitative analysis aimed to calculate the mean, frequency, and standard deviation. Meanwhile, responses to open-ended questions (a total of 3) underwent sentiment analysis conducted using the aforementioned software. Sentiment analysis falls within the domain of Natural Language Processing, focusing on the examination and interpretation of emotions, perspectives, attitudes, and tones expressed in written texts. The primary objective of sentiment analysis is to determine whether a text expresses viewpoints concerning a particular subject, product, or service (Dolianiti et al. 2018; Hutto & Gilbert 2014; Lius 2012; Mite-Baidal et al. 2018; Rao, Dhanya & Palathil 2020).

Sentiment analysis may offer valuable insights when applied to the examination of responses to open-ended questions regarding people's attitudes or opinions toward the use of digital technology in learning and education. Therefore, in this study, it is adopted to identify prevalent opinions among prospective support teachers within their responses, enabling the recognition of whether most responses exhibit a positive, negative, or neutral sentiment toward the role of digital technology in education.

The outcomes of the sentiment analysis may also serve as a valuable feedback mechanism for educators and policymakers in the field of education (Dolianiti et al. 2018; Zhao, Liu, & Xu 2015). This feedback may help tailor educational programs and shape policies that closely align with the perspectives and requirements of the educational community. For sentiment analysis, MAXQDA® utilizes a sentiment lexicon where each word is assigned to a sentiment score, encompassing positive, negative, slightly positive, slightly negative, neutral, or no reaction. MAXQDA® assesses each word within individual text fragments (e.g., sentences or paragraphs) against the lexicon and assigns the corresponding sentiment score. In cases where a word is not found in the lexicon, MAXQDA® seeks its lemma and, if located, employs the lemma's sentiment score. Two rules are applied to refine sentiment assessment. Firstly, in instances of negation, the scores of the subsequent three words are reversed. For example, "I was not very happy" would be classified as negative. Secondly, when modal verbs like "can" or "should" are used, sentiment is conveyed. Following the automatic coding process, it is also possible to aggregate sentiment scores across the collected data and generate visual representations or summary reports to gain insights into the sentiment trends within your qualitative data.

2.5 Results and Discussion

The sentiment analysis of the two open-ended questions ("How would you describe your approach regarding the use of digital and educational technologies in the teaching and learning process?" and "Do you think that in specific situations, determined by some disabilities, it could be difficult to use educational technologies?") categorized all the responses into six main groups:

- neutral, which implies that the text does not communicate either positive or negative attitude. It frequently denotes a lack of emotional or evaluative substance.
- Slightly positive: positive components are there in the text, although they are not fully conveyed.
- Slightly negative: like mildly positive, it signifies a moderate or subtle negative feeling. Negative components may be present in the text, but they are not forcefully expressed.
- Positive: text communicates a clear and obvious positive feeling.
- Negative sentiment: when a clear and obvious negative perspective or feeling is represented in the writing. It is frequently used to express criticism, displeasure, or disapproval.
- No reaction: when there is no obvious feeling represented in the text, this category is employed. It might be because of language utilized in the writing.

The results indicate a considerable variation in technological competence among participants. Furthermore, the responses exhibit numerous nuances that reflect the intricate nature of the digital education discourse, as illustrated in Tables 1 and 4.

In fact, the first analysis showed that the majority of the responded have a negative tendency to adopt educational technologies during their teaching in order to foster the learning process (34,15%) and that only the 5,69% show a positive attitude (Table 1).

Table 1. Frequency and percentage of segments coded for individual approaches to digital and educational technology use in teaching and learning.

	Frequency	%
Neutral	19	15,45
Slightly positive	16	13,01
Slightly negative	5	4,07
Positive	7	5,69

(continued)

Table 1. (*continued*)

	Frequency	%
Negative	42	34,15
No reaction	34	27,64
Total	123	100,00

Overall, there is a positive trend toward the use of digital technologies in education, with some participants expressing enthusiastic opinions, as highlighted in the following statements shared by the participants (Table 2).

Table 2. Examples of positive segments coded for individual approaches to digital and educational technology use in teaching and learning.

Category	Statements
Positive (5,69%)	"I'm a strong advocate for the use of digital technologies in education. They can provide personalized learning opportunities and enhance student engagement." "Digital tools have revolutionized my teaching. They allow me to create interactive and dynamic lessons tailored to individual student needs." "I firmly believe that technology is essential for preparing students for the future. It fosters critical thinking and problem-solving skills."
Slightly Positive (13,01%)	"I believe that digital technologies can enhance the learning experience. I've used them occasionally in my lessons and have seen positive results." "I have a basic level of competence with digital tools, and I find that they can make lessons more engaging for students." "Digital technologies have the potential to improve accessibility for all students, and I'm willing to explore their integration into my teaching methods."

However, there is also a significant presence of participants who show some uncertainty or fear about using technology or who claim to have no experience (Table 3).

On the contrary, at the question "Do you think that in specific situations, determined by some disabilities, it could be difficult to use educational technologies?", the software identified a portion of future support teachers' responses (33.33%) as lacking emotional or evaluative content. However, it also detected a moderate attitude, with 25.20% expressing a positive sentiment and 21.95% expressing a negative sentiment.

In addition, future support teachers who exhibited positive attitudes towards the question also expressed reservations regarding the adoption of educational technologies with students who have specific types of disabilities, including Autism Spectrum

Table 3. Examples of negative segments coded for individual approaches to digital and educational technology use in teaching and learning.

Category	Statements
Negative (34,15%)	"I have concerns about the overreliance on technology in education. It may lead to decreased face-to-face interactions and hinder social development." "I've seen instances where technology in the classroom has been distracting for students, and it can be challenging to maintain their focus." "I worry that the digital divide may exacerbate educational inequalities, with some students having access to technology while others do not."
Slightly Negative (4,07%)	"I'm somewhat hesitant about using digital technologies in the classroom because I'm not very tech-savvy. I'm concerned about technical difficulties." "While I see the potential benefits, I'm worried that the learning curve for using technology effectively may be too steep." "I've had some negative experiences with technology in the past, so I'm cautious about relying too heavily on it for teaching."

Table 4. Frequency and percentage of segments coded for challenges with educational technologies in specific disability contexts.

	Frequency	%
Neutral	41	33,33
Slightly positive	31	25,20
Slightly negative	27	21,95
Positive	14	11,38
Negative	5	4,07
No reaction	5	4,07
Total	123	100,00

Disorder, motor impairments, visual or auditory impairments, and intellectual disabilities, particularly in cases where the disabilities are severe in nature. This source of concern aligns with findings from prior studies. For instance, several researchers examined the challenges and reservations that teachers may face when integrating technology into classrooms for students with diverse impairments (Drijvers et al. 2017; Roblyer & Marshall 2002). These concerns often turn around questions of accessibility and the appropriateness of digital technologies for such students, especially those with severe impairments. These research outcomes underscore the importance of offering tailored support and training to address these apprehensions and ensure the implementation of inclusive digital education practices.

It is also noteworthy that the integration of educational technology in inclusive classrooms may encounter obstacles due to various factors. These factors were highlighted by future support teachers participating in this survey when responding to the question "In schools, which of the following factors may hinder teaching and learning practices with digital technologies?":

- inadequate digital competence among teachers (89.34%),
- insufficient availability of digital resources (72.13%),
- slow or unreliable internet connectivity (53.28%),
- limited financial resources (34.43%),
- scarce or nonexistent technical support (30.33%),
- space constraints within school premises (15.57%),
- limited digital competence among students (10.66%).

These factors highlighted the importance of having adequate digital resources and infrastructure for successful technology integration, the impact of network quality on digital learning experiences, the importance of having technical assistance available and financial limitations. These findings underscore the numerous challenges that may impede the effective use of digital technology in educational contexts, aligning with existing literature (Bouck, & Flanagan 2016; Hew, & Brush 2007; Inan, & Lowther 2010; Zhao et al. 2002).

Regarding factors related to teachers' competences, future support teachers participating in the investigation express confidence in their ability to prepare lessons by modifying or creating a range of digital resources (Mean = 4.51) and for conducting assessments or providing personal feedback and support to students (Mean = 4.54) (Table 5). However, it is noteworthy that despite this self-assessment, most of them feel that they have relatively limited competence in using educational technologies to:

- adapt their lessons and create digital resources to support their teaching practice (mean = 2.65),
- utilize virtual learning environments with students (mean = 2.48),
- design effective digital learning activities to engage students (mean = 2.56),
- employ digital and educational technologies to facilitate collaboration among students (mean = 2.48),
- assess students' learning (Table 6).

These findings suggest that future support teachers perceive a skill gap in specific parts of educational technology use in teaching and learning activities.

Table 5. Descriptive statistics: Area E and F "Pedagogy – tools, resources and classroom implementation"

	N	Mean	Std. Dev. (samp.)
Do you feel prepared to prepare lessons by modifying or creating a series of digital resources (e.g. presentations, images, audio or video)?	123	4,51	1,791
Do you feel prepared for classroom teaching practices using a variety of devices (e.g., interactive whiteboards, video projectors) and resources (e.g., online quizzes, mind maps, simulations)?	123	3,76	1,449
Do you feel prepared for assessment or personal feedback and support for students?	123	4,54	1,712
I know which online platforms to use to find digital educational resources	123	2,70	1,147
I am able to create digital resources to support my teaching practice	123	2,65	1,169
I am able to use virtual learning environments with students	123	2,48	1,107
I am able to use digital technologies for school-related communications	123	2,95	1,202
I am able to use digital and educational technologies to adapt my teaching practice to the needs of individual students	123	2,69	1,098
I am able to use digital and educational technologies to encourage students' creativity	123	2,76	1,157
I am able to design suitable digital learning activities to engage students	123	2,56	1,076
I know how to adopt digital and educational technologies to facilitate collaboration between students	123	2,48	1,039
I am able to encourage students to use digital and educational technologies for interdisciplinary projects	123	2,91	1,133

Table 6. Descriptive statistics: Area G "Assessment methods"

	N	Mean	Std. dev. (samp.)
They are able to adopt digital and educational technologies to assess students' skills	123	2,52	1,077
I know how to use digital and educational technologies to provide timely feedback to students	123	2,46	1,077
I am able to use digital technologies to help students reflect on their learning	123	2,45	1,053

3 Conclusions

The results of data analysis highlight a multifaceted relationship between future support teachers' digital skills and their attitudes toward the integration of ICT in inclusive educational settings. The survey also shows a heterogeneous reality. It can be observed that most respondents have reservations about using educational technology to improve learning outcomes, in line with the findings of Ertmer and Ottenbreit-Leftwich (2010), and Tondeur, Braak, Ertmer and Ottenbreit- Leftwich (2017), who noted that teachers' beliefs influence the adoption of technology in classrooms. Apprehensions about the use of assistive technology tools for students with disabilities confirm the sentiments found in the work of Okolo and Diedrich (2014), which underlined educators' uncertainties about the efficacy of technology integration for students with special needs. This hesitancy may be attributed to several impediments identified by the participants as insufficient resource, network reliability issues, lack of technical support and fiscal constraints, comparable to the barriers noted in Starks & Reich (2023). Moreover, participants revealed confidence in certain digital skills while acknowledging a deficit in others, suggesting a segmented expertise that necessitates targeted training. This is supported by the European Commission's (2017) DigCompEdu framework, which advocates for continuous professional development in digital competencies for educators. The overall narrative of the study highlights the complexity of ICT integration, emphasizing the importance of dedicated support and training to cultivate inclusive digital practices (Timotheou et al. 2023; Florian and Black- Hawkins 2011). Given the inherent connection between technology skills and the implementation of AI in education (Luckin et al. 2016), teachers should proactively engage in specialized training to effectively integrate and adapt AI-based tools to the needs of the classroom. This is critical, to close the skills gap and harness the potential of AI in enhancing personalized learning experiences (EC 2023; Xie et al. 2019). Considering the above, further studies will be conducted adapting the SELFIE questionnaire to investigate the relationship between future support teachers' opinions and perceptions of digital skills and the use of AI in inclusive teaching.

References

Bitner, N., Bitner, J.: Integrating technology into the classroom: Eight keys to success. J. Technol. Teach. Educ. **10**(1), 95–100 (2002)

Bocconi, S., Kampylis, P., Punie, Y.: Innovating learning: key elements for developing creative classrooms in Europe. JRC-IPTS (2012)

Bouck, E.C., Flanagan, S.M.: Factors influencing teachers' adoption of iPads for instructional purposes. J. Special Educ. Technol. **31**(2), 79–90 (2016)

Conrads, J., Rasmussen, M., Winters, N., Geniet, A., Langer L.: Overview and analysis of policies for the integration and effective Use of digital technologies in education Retrieved 12 September (2019)

Dolianiti, F., Iakovakis, D., Dias, S., Hadjileontiadou, S., Diniz, J., Hadjileontiadis, L.: Sentiment analysis techniques and applications in education: a survey. Commun. Comput. Inform. Sci. (2018). https://doi.org/10.1007/978-3-030-20954-4_31

Drijvers, P., Tacoma, S., Besamusca, A., Doorman, M.: Digital technology and special education needs: challenges and opportunities. In: Proceedings of the 12th International Congress on Mathematical Education (ICME-12), vol. 28, p. 194. Springer (2017)

Ertmer, P.A., Ottenbreit-Leftwich, A.T.: Teacher technology change: How knowledge, confidence, beliefs, and culture intersect. J. Res. Technol. Educ. **42**(3), 255–284 (2010)

European Commission. European Framework for the Digital Competence of Educators: DigCompEdu (2017)

European Commission: SELFIE (Self-reflection on Effective Learning by Fostering the use of Innovative Educational Technologies) - Digital Education Action Plan. Questo piano d'azione sottolinea l'importanza dell'auto-riflessione per l'apprendimento efficace e l'innovazione educativa (2018)

European Commission: White Paper on Artificial Intelligence - A European approach to excellence and trust. Publications Office of the European Union (2020)

Fengchun, M., Holmes, W.: Guidance for generative AI in education and research. UNESCO, p. 44 (2023)

Florian, L., Black-Hawkins, K.: Exploring inclusive pedagogy. Br. Edu. Res. J. **37**(5), 813–828 (2011)

Hew, K.F., Brush, T.: Integrating technology into K-12 teaching and learning: Current knowledge gaps and recommendations for future research. Educ. Technol. Res. Develop. **55**(3), 223–252 (2007)

Hutto, C.J., Gilbert, E.: VADER: a parsimonious rule-based model for sentiment analysis of social media text. In: Eighth International Conference on Weblogs and Social Media (ICWSM-14) (2014)

Inan, F.A., Lowther, D.L.: Factors affecting technology integration in K-12 classrooms: A path model. Educ. Technol. Res. Develop. **58**(2), 137–154 (2010)

Lim, C.-P., Zhao, Y., Tondeur, J., Chai, C.-S., Tsai C.-C.: Bridging the gap: Technology trends and use of technology in schools. Educ. Technol. Soc. **16**(2), 59–68 (2013)

Liu, B.: Sentiment Analysis and Opinion Mining (2012)

Luan, H., et al.: Challenges and future directions of big data and artificial intelligence in education. Front. Psychol. **11**, 580820 (2020)

Luckin, R., Holmes, W., Griffiths, M., Forcier, L.B.: Intelligence unleashed: an argument for AI in education. Pearson Education (2016)

Mite-Baidal, K., Delgado-Vera, C., Solis-Avilés, E., Espinoza, A., Ortiz-Zambrano, J., Varela-Tapia, E.: Sentiment analysis in education domain: a systematic literature review, pp. 285–297 (2018). https://doi.org/10.1007/978-3-030-00940-3_21

OECDTALIS 2018 Results (Volume I): Teachers and School Leaders as Lifelong Learners (2019)

Okolo, C.M., Diedrich, J.: Twenty-five years later: How is technology used in the education of students with disabilities? Results of a statewide study. J. Spec. Educ. Technol. **29**(1), 1–20 (2014)

Rao, M., Dhanya, M., Palathil, A.: An analysis of customer sentiments towards education technology app: a text mining approach. Soc. Sci. Res. Netw. (2020). https://doi.org/10.2139/ssrn.3563577

Redecker, C:. European framework for the digital competence of educators: DigCompEdu. Edited by Yves Punie (2017)

Roblyer, M.D., Marshall, J.C.: Predicting the integration of technology in preservice teacher education: A model using the theory of planned behavior. J. Res. Comput. Educ. **34**(3), 263–276 (2002)

Rogers, A.: The base of the iceberg: informal learning and its impact on formal and non- formal learning. Barbara Budrich Publishers (2014)

Sánchez-Prieto, J.C., Olmos-Migueláñez, S., García-Peñalvo, F.J.: MLearning and pre- service teachers: an assessment of the behavioral intention using an expanded TAM model. Comput. Hum. Behav. **72**, 644–654 (2017)

Selwyn, N.: Education and technology: key issues and debates. Bloomsbury Academic (2017)

Skryabin, M., Zhang, J., Liu, L., Zhang, D.: How the ICT development level and usage influence student achievement in reading, mathematics, and science. Comput. Educ. **85**, 49–58 (2015)

Software, V.E.R.B.I.: MAXQDA. VERBI Software Berlin (2019)

Starks, A.C., Reich, S.M.: What about special ed?: barriers and enablers for teaching with technology in special education. Comput. Educ. **193**, 104665 (2023)

Timotheou, S., Miliou, O., Dimitriadis, Y., et al.: Impacts of digital technologies on education and factors influencing schools' digital capacity and transformation: a literature review. Educ. Inf. Technol. **28**, 6695–6726 (2023). https://doi.org/10.1007/s10639-022-11431-8

Tondeur, J., Van Braak, J., Ertmer, P.A., Ottenbreit-Leftwich, A.: Understanding the relationship between teachers' pedagogical beliefs and technology use in education: a systematic review of qualitative evidence. Educ. Tech. Res. Dev. **65**, 555–575 (2017)

UN. United Nations (2015). Resolution adopted by the General Assembly. Transforming our world: the 2030 agenda for sustainable development. A/RES/70/1, 25 September 2015. https://www.un.org/en/development/desa/population/migration/generalassembly/docs/globalcompact/A_RES_70_1_E.pdf (ver. 15.04.2021)

Wang, Y., Liu, C., Tu, Y.F.: Factors affecting the adoption of AI-based applications in higher education. Educ. Technol. Soc. **24**(3), 116–129 (2021)

Xie, Y., Ke, F., Sharma, P.: The effect of peer collaboration on children's problem-solving ability during game-based learning. Interact. Learn. Environ. **27**(5–6), 690–704 (2019)

Zhang, K., Aslan, A.B.: AI technologies for education: Recent research & future directions. Comput. Educ. Artific. Intell. **2**, 100025 (2021)

Zhang, L., Basham, J.D., Yang, S.: Understanding the implementation of personalized learning: a research synthesis. Educ. Res. Rev. **31**, 100339 (2020)

Zhao, J., Liu, K., Xu, L.: Book review: sentiment analysis: mining opinions, sentiments, and emotions by Bing Liu. Comput. Linguist. **42**, 595–598 (2015). https://doi.org/10.1162/COLI_r_00259

Zhao, Y., Cziko, G.A.: Teacher adoption of technology: a perceptual-control-theory perspective. J. Technol. Teach. Educ. **9**(1), 5–30 (2001)

Zhao, Y., Pugh, K., Sheldon, S., Byers, J.L.: Conditions for classroom technology innovations. Teachers College Record **104**(3), 482–515 (2002)

Future Support Teachers' Intentions to Adopt and Use Technology with Students with ASD

Emanuela Zappalà[(⊠)] [ID], Ilaria Viola[ID], and Paola Aiello[ID]

University of Salerno, Salerno, Italy
ezappala@unisa.it

Abstract. Artificial intelligence (AI) and technology have the potential to transform the teaching-learning process when adopted in inclusive classroom also attended students with Autism Spectrum Disorder (ASD). However, there is a lack of research on teachers' intentions to adopt and use AI and technology with students with ASD. Therefore, this study explores the factors that may influence teachers' intentions to use AI and technology with students with ASD by adopting the Unified Theory of Acceptance and Use of Technology.

Keywords: Artificial Intelligence · ICT · Teachers' intentions · Autism Spectrum Disorder · Inclusion

1 Introduction

Autism Spectrum Disorders (ASD) represent a complex neurodevelopmental condition characterized by a spectrum of manifestations in social communication, interaction, and repetitive behaviors. Due to its heterogeneity in severity levels and the variety of symptoms, selecting and implementing effective practices for teachers to promote the inclusion of students with ASD in the classroom is a challenging task.

The use of Artificial Intelligence (AI) and educational technologies (such as, Information and Communication Technologies and Assistive Technology) has emerged as an appealing strategy for dealing with some of these issues. In recent decades, numerous systematic reviews and meta-analyses have investigated the efficacy of educational technologies in facilitating the comprehensive development and learning processes of students with ASD (Hume et al. 2021; Odom et al. 2015; Zervogianni et al. 2020). Overall, these reviews have highlighted that the use of some technologies may address the diverse needs of these students, enhancing and fostering the acquisition of social, emotional, academic skills, or developing adaptive behaviors (Bolte, Golan, Goodwin, & Zwaigenbaum 2010).

The article is the result of a collaboration between the authors. However, Emanuela Zappalà wrote paragraphs 1 "Introduction" and 2 "Teachers' readiness to adopt ICT and AI in school"; Ilaria Viola wrote paragraphs 2 "Methodology" and "Conclusion"; Paola Aiello is the scientific coordinator of the research.

Moreover, AI may also offer a range of potential benefits, which include tailored interventions, enhanced communication, and personalized learning experiences (Mazurek et al. 2012). For example, AI-powered tools may be used to provide personalized instruction, support social communication skills, and help students to self- manage their behavior (Barua et al. 2022). Researchers explored the application of AI in improving social skills and academic outcomes for children with ASD. Cao et al. (2019) employed sensory inputs like facial expressions, body movements, and voice recordings to develop AI-driven robots that analyzed ASD peoples' behavior and engagement during therapy sessions, demonstrating AI's potential for enhancing social interaction. Sanghvi et al. (2011) expanded this approach by employing postural expressions in activities such as chess to assess engagement levels in ASD children, suggesting the integration of AI and affect recognition systems for real-world support. Cha et al. (2021) employed audio recordings and machine learning to evaluate emotions and engagement during interactions with robots. Additionally, Rudovic et al. (2018) developed a personalized deep learning model combining various data sources, achieving human-level accuracy in engagement assessment. Integration of speech recognition and AI in voice-interactive educational robots further demonstrated AI's capacity to enhance educational support and social interaction for children with ASD, emphasizing AI's role in addressing their unique challenges. Through sophisticated algorithms and data-driven insights, AI systems can identify learning patterns, strengths, and weaknesses, allowing for the creation of customized learning pathways. Such tailored interventions not only optimize the learning experience but also accommodate the diverse learning profiles often observed among individuals with ASD (Ramdoss et al. 2012). Furthermore, tailored interventions constitute one of the main advantages of AI-powered tools. These technologies may adapt instructional content and strategies to suit the individualized needs and preferences of students with ASD (such as, Proloquo2Go or LAMP Words for Autism which foster communication skills, for example).

For these reasons, AI and technology may have the educational potential to address the challenges encountered by students with ASD and offer opportunities to empower teachers and enhance personalized and individualized teaching practices. Despite that, there are some concerns about AI's integration in inclusive schools. In fact, even if, the use of AI and technologies to enhance learning has exponentially grown over the last decade, partly prompted by the COVID-19 school closures, there is a lack of information about AI's real influence on learning outcomes and its potential to broaden our understanding of successful learning processes (Miao et al. 2021). In addition, several studies investigated the design and effectiveness of ICT, but limited research explored the factors influencing teachers' intentions to use of both ICT and AI tools in special and inclusive education. Furthermore, for teachers to effectively engage with AI, they should acquire new skills and participate in appropriate professional development initiatives to employ its potential.

Thus, considering that there is a strong correlation between teachers' intention to use educational technologies and AI to foster inclusion and their intentions and actions (Aiello, Sharma 2018; Ajzen & Fishbein, 1980; Fishbein and Ajzen 1975; Vidal-Hall, Flewitt, Wyse 2020), the study aims at investigating the intentions of future support teachers to adopt and use the technology by using the unified theory of acceptance and use of technology (UTAUT; Venkatesh et al. 2003).

2 Teachers' Readiness to Adopt ICT and AI in School

Teachers play a central role in integrating ICT and AI into the learning process, fostering meaningful learning situations and employing specific *educational mediators* to engage students (Goussot 2014). Teachers' readiness to adopt AI in educational settings may be influenced by multiple factors:

– teachers' preparedness and perceptions to incorporate AI tools into their teaching practices (Al-Gahtani 2022; Liaw & Huang 2021; McCarthy & O'Connor's 2022);
– the availability of technological infrastructure, the adequacy of training, the perceived utility of AI tools, and contextual considerations (Chen, & Zhang 2022; Liaw & Huang 2021);
– the requisite allocation of resources and support (Elhoweris & Al-Gahtani 2023; Liaw & Huang 2021).

It emphasizes the multifaceted nature of AI adoption, strictly connected with various factors which emphasize the need for tailored strategies, specialized training, and dedicated support to facilitate the seamless assimilation of AI technologies into pedagogical contexts (Al-Gahtani 2022; Chen, & Zhang 2022; Elhoweris & Al- Gahtani 2023; Liaw & Huang 2021; McCarthy & O'Connor's 2022).

One notable aspect the reviews mention is the lack of standardized tools for assessing teachers' readiness to adopt AI and educational technologies in classrooms with students with ASD. This gap in research is significant since prior investigations have highlighted the strong link between technological skills and AI adoption. Moreover, technological skills are often considered essential for fully employing the educational potential of AI in educational contexts (Sánchez-Prieto et al. 2017; Wang, Liu, Tu 2021).

A theory whose tool may address this purpose is the UTAUT (Venkatesh et al. 2003). It is a theoretical model that identifies the fundamental factors influencing technology acceptance and utilization. According to this theory, the actual use of technology is driven by an individual's behavioral intention which may be impacted by five main constructs: performance expectancy, effort expectancy, social influence, facilitating conditions and intrinsic value. *Performance expectancy* may be defined as the perception that using a system would improve work performance. It is a strong predictor of usage intention and is important in both voluntary as well as mandatory situations (Zhou et al. 2010). The second, *effort expectation*, assess how easy it is to use a system, although its significance tends to fade after a while (Gupta et al. 2008). The third construct, *social influence*, is conceived as the perception of others' beliefs about using the system, is particularly influential when technology usage is when technology utilization is compulsory (Venkatesh et al. 2003). The fourth one, *facilitating conditions*, is related to the belief that organizational and technical support have a direct influence on use intention (Venkatesh et al. 2003); whereas the last construct, *intrinsic value*, might include the personal satisfaction or enjoyment derived from using technology and it may influence the willingness and readiness to use technology (Khechine et al. 2020).

All these factors may have a direct impact on one's intention to adopt technology and, eventually, their actual use. Furthermore, UTAUT emphasizes that individual traits might alter these factors as gender may affects all of them, experience may moderate relationships related to effort expectancy, social influence and facilitating conditions, and voluntariness of use may modify the link between social influence and behavioral intention (Venkatesh et al. 2003). UTAUT has greatly contributed to technology acceptance research by combining several theories of technology acceptance and developing a strong prediction model that accounts for 70% of the variance in use intention (Venkatesh et al. 2003). In terms of predictive power, it improves prior models such as the Technology Acceptance Model (TAM) (Davis 1993).

Building upon the analysis of Venkatesh (2022) and the research group of Han and Conti (2020) and Hajjar et al. (2021), the UTAUT may be adopted as a theoretical foundation on technology adoption with students with ASD, particularly in the context of AI tools but also offer organizations valuable insights into promoting the adoption of AI tools in inclusive classrooms.

3   Methodology

3.1   Objective

The purpose of this study is that of investigating the intentions of future support teachers to adopt and use the technology. Specifically, the aim is to detect which independent variables affect intentional behavior in using technologies with students with ASD.

3.2   Sample and Procedure

The study uses a convenience sample of 122 potential upper secondary support teachers enrolled in the Specialization course for educational support activities for students with disabilities at the University of Salerno (a.y. 2021/2022). Data are gathered after a training course leading to the acquisition of the warrant to work as learning support teachers in upper secondary school, that is divided into different modules totaling 750 h, composed of lectures, tutorials, workshops, on-site teaching practice and tutorials. In particular, the questionnaire is administered at the beginning of the first class of "Information and Communication Technologies" module which lasts 75 h.

All the course participants (N = 122) were asked to complete the questionnaire voluntarily and anonymously.

3.3   The Data Collection Tool

An online questionnaire with two parts was used to collect the data: a section dedicated to detecting the gender and age of the participants, and a second part of 41 items aimed at investigating the intentions of future support teachers to adopt and use a new technology. The scale was structured from the version by Venkatesj et al. (2003), Oye et al. (2012) by Khechine et al. (2020), and is divided into 6 areas:

Performance Expectancy (PE) according to the UTAUT model of Venkatesh and coworkers (Venkatesh et al. 2003) is a key component in understanding how individuals assess the impact of technology use on their job performance. This expectation is measured through a series of statements ranging from PE1 to PE10, which explore in detail users' perceptions of the contribution technology can make to improving their work activities. People tend to adopt the use of technologies that they perceive as helpful in increasing their efficiency and productivity.

In parallel, Effort Expectancy (EE) is examined through items ranging from EE1 to EE8 and relates to perceived ease of using technology. This dimension focuses on how intuitive and effortless it is to adopt and use a new technology. If users view a technology as easy to use, they are more likely to develop a positive intention toward its use (Venkatesh et al. 2003).

The Social Influence construct, investigated from SI1 to SI5, captures the extent to which perceptions of pressure or support from peers, superiors, and significant individuals can influence a user's decision to adopt a technology. This dimension is based on the theory that individuals are often influenced in their actions by the opinions and attitudes of the people around them (Venkatesh et al. 2003).

Facilitating Conditions, explored by items FC1 to FC5, refer to individuals' perceptions of the existence of appropriate organizational and infrastructural support that can facilitate the effective use of a technology. If users believe that the organization provides the necessary support, such as training or technical assistance, they are more likely to use the technology (Venkatesh et al. 2003).

In technology adoption models, intention plays a critical role as an antecedent to action. Fishbein and Ajzen (1975) emphasized the importance of measuring behavioral intention in predicting the execution of a voluntary action, unless the intention itself changes prior to the execution of the action. This behavioral intention is examined by items BI1 to BI5.

Finally, the construct of intrinsic value is described as the feeling of pleasure and interest in performing an activity, a dimension that traces intrinsic motivation. This concept, as explained by Wigfield and Eccles (2000), refers to the interest and enjoyment derived from performing a task for its own sake, rather than to achieve external goals or rewards, and may be a predisposing factor for the adoption and continued use of a technology, regardless of perceived performance or benefits. Intrinsic value is investigated by items V1 to V4 (Khechine et al. 2020).

These items were rated on a 5-point Likert-type scale (strongly disagree to strongly agree).

3.4 Analysis

First, a descriptive data analysis was carried out with respect to age and sex and subtests. For the purpose of detecting the reliability and internal consistency of the test, Cronbach's Alpha was calculated. Finally, to investigate the impact between the independent variables (PE, EE, IS, FC, VI) and intentional behavior (BI) a multiple linear regression analysis was conducted with SPSS.

3.5 Results

The descriptive analysis regarding age shows that 40.2% are aged 30–39 as shown in Table 1. Instead, 81% of the sample is female (N = 99) and 18.9% is male (N = 23).

Table 1. Age descriptive analysis

Age	Frequency	Percentage
25	1	,8
25–29	12	9,8
30–39	49	40,2
40–49	45	36,9
50–59	15	12,3

Data analysis shows that participants have high intrinsic value (Mean = 4.00) in using ICT, and followed by intentional behavior (Mean = 3.68) and high Facilitating Conditions (Mean = 3.67), see Table 2.

Internal consistency was tested with Cronbach's alpha (α) obtained from the SPSS software, as shown in Table 2, the alpha values for all measurement instruments were satisfactory, above .7 as recommended by Nunnally (1978) and Gerbing and Anderson (1988). Only the EE and SI subtests have values that fall within an average range.

Table 2. Descriptive subtest and reliability analysis

	Mean	Standard deviation	Alfa of Cronbach
BI	3,68	,637	,829
PE	3,56	,647	,884
EE	3,41	,489	,671
SI	3,47	,577	,492
FC	3,67	,674	,913
IV	4,00	,815	,950

Multiple regression analysis shows that the path coefficient of the relationship between facilitating conditions and behavioral intention is significant and positive (β = .352, t = 4.583, p < .05), see Table 3. The effect of facilitating conditions on use behavior is positive (β = .352,) and significant (t = 4.583, p < .01). The effect of intrinsic value on behavioral intention is positive (β =,358) and significant (t = 6.883, p < .01).

Consistent with the literature (Khechine et al. 2020), facilitating conditions and intrinsic value were found to be the main drivers of behavioral intention with an explained variance of 70.3%.

Table 3. Structural model results

	Dependent variables	behavioral intention	
		$R2 = ,703$	
Independent variables	β	t	Sig.
PE	−,047	−,646	,520
EE	,164	1,707	,091
SI	,074	1,072	,286
FC	,352	4,583	,000
IV	,358	6,883	,000

4 Conclusions

The UTAUT is a theoretical framework used to understand the adoption and use of technology which may be considered valuable for understanding the acceptance and usability of ICT in special and inclusive fields. In line with that, the investigation conducted with future support teachers demonstrate that the use of technology in classrooms, also attended by students with ASD, is significantly influenced by the availability of facilitating conditions and the intrinsic value they find in the use of technology. In fact, results show a positive and significant effect of facilitating conditions on use behavior; it means that future support teachers are more likely to engage and adopt educational technologies with students with ASD when the perceive that support, infrastructure and resources required to use it are available.

Moreover, results show intrinsic value is one of the main drivers of behavioral intention, which suggests that future support teachers find personal satisfaction or pleasure in the use of technology, so it seems that they are positively inclined to use it with students with ASD. This intrinsic motivation may be a positive factor in encouraging the use of technology for learning, communication and skills development. The research findings reinforce the idea that both practical support for technology use and intrinsic value are crucial in determining whether teachers will choose to use and benefit from technology tools in education.

Since Venkatesh et al. (2003) identify further moderating factors (gender, age, experience) that may influence the relationships between core constructs of the model and users' behavioral intentions, further investigation in the field of special and inclusive education are necessary. In fact, Venkatesh et al. (2003) suggest that younger users might be more influenced by social influence (the degree to which an individual perceives that important others believe they should use the new system) and facilitating conditions (the degree to which an individual believes that an organizational and technical infrastructure exists to support the use of the system) compared to older users. Likewise, previous experience with technology may also moderate the relationship between UTAUT constructs and behavioral intention.

In practical terms, the fact that intrinsic value was the most significant predictor of individual's intention to use technology may imply that more attention should be devoted to the design and features of such a system. Designers may aim to reduce monotony and harness features that users may find enjoyable, entertaining, and engaging. Decision-makers tasked with selecting a device may also consider not only its functionality but also its capacity to provide enjoyment, amusement, and foster interest. Consequently, the results emphasize the importance of promoting the intrinsic value to support future teachers in using educational technologies. Furthermore, they suggest creating more favorable conditions for the increased adoption of technologies in the classroom.

Lastly, considering that gender, age, and previous experiences may facilitate the interaction between the UTAUT constructs and intentional behavior (Venkatesh et al. 2003), further studies will be conducted to replicate the investigation, expanding the sample to explore how specific individual characteristics of future support teachers may influence the variables.

References

Aiello, P., Sharma, U.: Improving intentions to teach in inclusive classrooms: the impact of teacher education courses on future Learning Support Teachers. Form@ re-Open Journal per la formazione in rete **18**(1), 207–219 (2018)

Ajzen, I., Fishbein, M.: Understanding attitudes and predicting social behavior. Prentice Hall, Englewood Cliffs, NJ (1980)

Al-Gahtani, S.S.: Teachers' readiness to use artificial intelligence in their teaching practices. Educ. Inf. Technol. **27**(1), 1–19 (2022)

Barua, P.D., et al.: Artificial intelligence enabled personalised assistive tools to enhance education of children with neurodevelopmental disorders—a review. Int. J. Environ. Res. Public Health **19**(3), 1192 (2022)

Bolte, S., Golan, O., Goodwin, M.S., Zwaigenbaum, L.: What can innovative technologies do for Autism Spectrum Disorders? Autism **14**(3), 155–159 (2010)

Cao, H.L., et al.: Robot-enhanced therapy: development and validation of supervised autonomous robotic system for autism spectrum disorders therapy. IEEE Robot. Autom. Mag. **26**(2), 49–58 (2019)

Cha, I., Kim, S.I., Hong, H., Yoo, H., Lim, Y.K.: Exploring the use of a voice-based conversational agent to empower adolescents with autism spectrum disorder. In: Proceedings of the 2021 CHI Conference on Human Factors in Computing Systems, pp. 1–15 (2021)

Chen, B., Zhang, Z.: Factors influencing teachers' adoption of artificial intelligence in education: a systematic review. Comput. Educ. **178**, 104673 (2022)

Davis, F.D.: User acceptance of information technology: system characteristics, user perceptions and behavioral impacts. Int. J. Man Mach. Stud. **38**, 475–487 (1993)

Elhoweris, H., Al-Gahtani, S.S.: Teachers' perceptions of the benefits and challenges of using artificial intelligence in education. Educ. Inf. Technol. **28**(2), 1–20 (2023)

European Commission: White Paper on Artificial Intelligence - A European approach

Fishbein, M., Ajzen, I.: Belief, attitude, intention and behavior: An introduction to theory and research. Addison-Wesley, Reading, MA (1975)

Gerbing, D.W., Anderson, J.C.: An updated paradigm for scale development incorporating unidimensionality and its assessment. J. Mark. Res. **25**(2), 186–192 (1988)

Goussot, A.: Per una critica pedagogica della ragione differenzialistica tra "bisogni educativi speciali" e processi inclusivi. In: Pedagogia speciale e "BES". Spunti per una riflessione critica verso la scuola inclusiva (2014)

Gupta, B., Dasgupta, S., Gupta, A.: Adoption of ICT in a government organization in a developing country: an empirical study. J. Strateg. Inf. Syst. **17**(2), 140–154 (2008)

Hajjar, M.F., Alharbi, S.T., Ghabban, F.M.: Usability evaluation and user acceptance of mobile applications for Saudi autistic children. Int. J. Interact. Mobile Technol. **15**(7) (2021)

Han, J., Conti, D.: The use of UTAUT and post acceptance models to investigate the attitude towards a telepresence robot in an educational setting. Robotics **9**(2), 34 (2020)

Hume, K., et al.: Evidence-based practices for children, youth, and young adults with autism: third generation review. J. Autism Develop. Disorders 0123456789 (2021). https://doi.org/10.1007/s10803-020-04844-2

Khechine, H., Raymond, B., Augier, M.: The adoption of a social learning system: intrinsic value in the UTAUT model. Br. J. Edu. Technol. **51**(6), 2306–2325 (2020)

Liaw, S.S., Huang, H.M.: Teachers' intention to use artificial intelligence in education: a meta-analysis. Educ. Tech. Res. Dev. **69**(6), 3611–3634 (2021)

Mazurek, M.O., Shattuck, P.T., Wagner, M., Cooper, B.P.: Prevalence and correlates of screen-based media use among youths with autism spectrum disorders. J. Autism Dev. Disord. **42**, 1757–1767 (2012)

McCarthy, E., O'Connor, R.: Teachers' attitudes towards the use of artificial intelligence in education: a systematic review. Educ. Technol. Soc. **25**(3), 31–48 (2022)

Miao, F., Holmes, W., Huang, R., Zhang, H.: AI and education: a guidance for policymakers. UNESCO Publishing (2021)

Nunnally, J.C.: An overview of psychological measurement. Clinical diagnosis of mental disorders: a handbook, pp. 97–146 (1978)

Odom, S.L., et al.: Technology-aided interventions and instruction for adolescents with autism spectrum disorder. J. Autism Dev. Disord. **45**(12), 3805–3819 (2015)

Oye, N.D., Iahad, N., Nor, Z.A.R.: The impact of UTAUT model and ICT theoretical framework on university academic staff: focus on Adamawa State University, Nigeria. Int. J. Comput. Technol. **2**(2), 102–111 (2012)

Ramdoss, S., Machalicek, W., Rispoli, M., Mulloy, A., Lang, R., O'Reilly, M.: Computer- based interventions to improve social and emotional skills in individuals with autism spectrum disorders: a systematic review. Dev. Neurorehabil. **15**(2), 119–135 (2012)

Rudovic, O., Lee, J., Dai, M., Schuller, B., Picard, R.W.: Personalized machine learning for robot perception of affect and engagement in autism therapy. Sci. Robot. **3**(19) (2018)

Sánchez-Prieto, J.C., Olmos-Migueláñez, S., García-Peñalvo, F.J.: MLearning and pre- service teachers: an assessment of the behavioral intention using an expanded TAM model. Comput. Hum. Behav. **72**, 644–654 (2017)

Sanghvi, J., Castellano, G., Leite, I., Pereira, A., McOwan, P.W, Paiva, A.: Automatic analysis of effective postures and body motion to detect engagement with a game companion. In: 2011 6th ACM/IEEE International Conference on Human-Robot Interaction(HRI), pp. 305–311 (2011). to excellence and trust. Publications Office of the European Union (2020)

Venkatesh, V.: Adoption and use of AI tools: a research agenda grounded in UTAUT. Ann. Oper. Res. 1–12 (2022)

Venkatesh, V., Morris, M.G., Davis, G.B., Davis, F.D.: User acceptance of information technology: toward a unified view. MIS Q. **27**(3), 425–478 (2003)

Vidal-Hall, C., Flewitt, R., Wyse, D.: Early childhood practitioner beliefs about digital media: integrating technology into a child-centred classroom environment. Eur. Early Child. Educ. Res. J. **28**, 167–181 (2020)

Wang, Y., Liu, C., Tu, Y.F.: Factors affecting the adoption of AI-based applications in higher education. Educ. Technol. Soc. **24**(3), 116–129 (2021)

Wigfield, A., Eccles, J.S.: Expectancy–value theory of achievement motivation. Contemp. Educ. Psychol. **25**(1), 68–81 (2000)

Zervogianni, V., et al.: A framework of evidence-based practice for digital support, co- developed with and for the autism community. Autism **24**(6), 1411–1422 (2020). https://doi.org/10.1177/1362361319898331

Zhou, T., Lu, Y.B., Wang, B.: Integrating TTF and UTAUT to explain mobile banking user adoption. Comput. Hum. Behav. **26**(4), 760–767 (2010)

Generative AI-Driven Digital Assistance for E-Learning: A Novel Paradigm for Personalized Recommendations

Ha X. Son[1(✉)], Triet M. Nguyen[2(✉)], Hong K. Vo[2], Khoa T. Dang[2], Khiem H. Gia[2], Nam B. Tran[2], Bang L. Khanh[2], and Ngan T. K. Nguyen[3]

[1] RMIT University, SGS campus, Ho Chi Minh city, Vietnam
ha.son@rmit.edu.vn
[2] FPT University, Can Tho city, Vietnam
trietnm3@fe.edu.vn
[3] FPT Polytechnic, Can Tho city, Vietnam

Abstract. With the continuous proliferation of E-Learning platforms, the demand for intelligent and adaptive systems to guide learners through vast content repositories has grown. Drawing inspiration from recent advancements in recommendation systems, particularly within digital library contexts, this paper presents an innovative approach using Generative AI to define digital assistance in E-Learning environments. Unlike traditional recommendation systems, which suggest resources based on explicit patterns or user metadata, our Generative AI model dynamically crafts personalized content based on learners' preferences and progress. We juxtapose our methods with prevalent deep learning-based recommendation systems as discussed in prior research. The novelty of our approach lies in the amalgamation of generative algorithms with personalized recommendation, thereby offering a dynamic, real-time, and context-aware learning guide. The paper elaborates on the model's architecture, its performance metrics in comparison to existing methods, and its implications for the future of digital education. Through this study, we hope to pave the way for more intuitive, adaptive, and responsive E-Learning experiences.

Keywords: Generative AI · Digital Assistant · Digital Assistance · Personalized Recommendations · Content Personalization · Adaptive Learning Systems · Digital Library

1 Introduction

The dawn of the digital age has brought with it a paradigm shift in how we perceive and interact with libraries. What were once static, physical collections of books and articles have now expanded to become dynamic, ever-evolving repositories accessible from anywhere in the world [6,11]. This transformation, however profound, introduces a complexity of navigating an often overwhelming breadth

© The Author(s), under exclusive license to Springer Nature Switzerland AG 2024
F. Palomba and C. Gravino (Eds.): WAILS 2024, LNCS 14545, pp. 89–98, 2024.
https://doi.org/10.1007/978-3-031-57402-3_10

and depth of digital content. Enter recommendation systems, a beacon in this vast digital expanse. [6] innovatively employs graph-based structures to map and suggest content, optimizing navigability and discoverability. Concurrently, [11] emphasizes on a quality-driven approach, ensuring that the disseminated information within university digital libraries remains not only relevant but also of high caliber. Together, these systems play a cardinal role in refining the digital library user experience, adapting to the users' needs, and guiding them to pertinent information seamlessly.

In recent years, the domain of recommendation systems has witnessed an upsurge of innovations, prominently spearheaded by the integration of deep learning methodologies [2,3,13]. The underlying rationale is evident: the ability of deep learning to discern intricate patterns and predict user preferences with remarkable accuracy. For instance, [3] goes beyond conventional recommendation algorithms and leverages deep learning to align recommendations with users' specific research interests. Delving into the linguistic aspect, [2] stands out by bridging the cultural and language gaps prevalent in digital libraries. Echoing and amplifying these innovative strides, [13] provides an encompassing overview. The systematic review unfurls the vast landscape of deep learning-driven recommendation systems, shedding light on their transformative potential and charting the course for future endeavors in this vibrant field.

In the ever-advancing landscape of digital recommendations, there emerges an intersection of semantic understanding and intricate user metadata, fostering a richer, more personalized content discovery experience. At the heart of this shift lies the innovative work of [14], which unveils pattern-based hybrid models. These models not only tap into the inherent semantics of content but also knit together diverse data sources, weaving a richer tapestry of recommendations. Parallelly, [15] champions a unique approach, fusing user preferences at a granular level. By interlinking multiple facets of a user's digital footprint, the resulting recommendations encapsulate a balance of what users explicitly express and implicitly hint at. Collectively, these advances underscore an era where recommendations transition from being mere suggestions to well-crafted narratives, aligning seamlessly with a user's academic pursuits, personal inclinations, and ever-evolving interests.

Traditional libraries, though repositories of knowledge, often lack the dynamism to cater to the specific, evolving needs of academia. This gap is addressed by innovative systems that tailor recommendations to the nuanced interests of scholars [3,10,12]. In this regard, [5] pioneers an approach that is not only data-driven but also subject-specific. By homing in on specific reading subjects and harnessing the power of robust data analysis, the system ensures that recommendations mirror the academic pulse of users. Meanwhile, [1] introduces a fascinating dimension, weaving the social fabric into the recommendation engine. By integrating cues from social networks, the system hints at the transformative potential of community-curated and community-driven content suggestions, fostering collaborative learning and shared academic exploration.

Our research delves into the integration of Generative AI mechanisms into recommendation systems, situating our exploration amidst the backdrop of foundational studies. Drawing inspiration from pivotal literature, we ventured into the uncharted waters of Generative AI-driven methodologies in E-Learning digital assistance. The subsequent sections of this manuscript detail our innovative proof-of-concept, which harnesses the power of Generative AI for tailored content recommendations. Through rigorous evaluation involving students from FPT University, we aim to gauge its efficacy and potential to revolutionize digital learning experiences. Our findings provide insights into how this synergetic fusion of traditional algorithms and contemporary generative techniques can usher in a new era in E-Learning, attuned to the modern scholar's needs.

2 Related Work

The field of recommendation systems, especially as applied to digital libraries and educational resources, has witnessed prolific advancements over the last two decades. This section surveys the key contributions and situates our investigation within the broader landscape, emphasizing the novelty of our Generative AI-driven approach.

Early works primarily leveraged traditional algorithms to enhance user experience in digital libraries. For instance, [6] introduced a graph-based recommendation system, while [11] focused on quality-centric recommendations to better disseminate information within university digital libraries. On the other hand, [4] experimented with Hebbian algorithms for recommendations, and [7] employed association rules based on user profiles.

The subsequent wave of research was marked by the integration of deep learning methods. [3] harnessed users' research interests through deep learning, and [2] concentrated on metadata-driven recommendations for both Arabic and English languages. In a broader scope, [13] provided a comprehensive survey on the role of deep learning in recommendation systems.

Some research works delved into the intricacies of personalizing recommendations. For example, [9] designed a recommendation system using Deep Belief Networks, while [15] fused user preferences to construct smarter libraries. Further emphasizing the importance of personalization in academic settings, [8] proposed a recommendation system tailored for academic research papers.

A noteworthy paradigm shift was observed with the advent of hybrid models, exemplified by [14], which combined semantic relationships in a pattern-based recommendation approach. Concurrently, the role of social networks and community-driven insights became evident in works such as [1], hinting at the potential of collaborative filtering and socially infused recommendation mechanisms. Against this backdrop, our paper presents a distinct approach, leveraging the potential of Generative AI. While the aforementioned studies have undeniably enriched the domain of recommendation systems, none have hitherto explored the capabilities and nuances of Generative AI in crafting personalized E-Learning digital assistance. Our work seeks to bridge this gap, synergizing established methodologies with the unparalleled potential of Generative

AI, thereby heralding a novel paradigm for personalized recommendations in E-Learning.

3   Approach

In our study, several components play crucial roles in enabling an advanced recommendation system for the E-learning platform (see Fig. 1). The figure showcases the foundational role of the *Material for Education* that feeds into the *Data Preparation Process for Digital Library*. This process integrates tools like *Chroma, Langchain, Chat GPT*, and collects *Metadata*. The data, once processed, is stored in *MongoDB* and managed using the *NodeJS* programming environment. All this culminates in the creation of the *Digital Library*, which prepares prompts for the final component: the *E-learning System*, designed to provide an array of personalized educational services. This section elaborates on each of these components in detail.

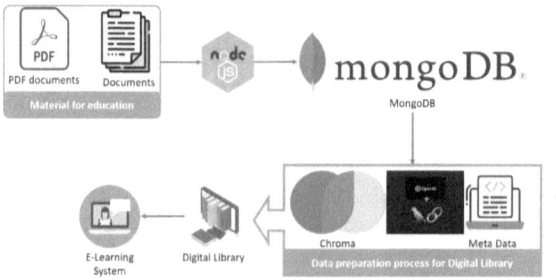

Fig. 1. Architecture of the Generative AI-Driven Digital Assistance System for E-Learning

Material for Education. The foundation of any effective E-learning system is the quality and relevance of educational material. This encompasses textual content, multimedia resources, interactive quizzes, and more, catering to diverse learning modalities and ensuring comprehensive coverage of topics. The materials serve as the primary knowledge base, which the recommendation system taps into, ensuring that learners receive content that aligns with their needs and preferences.

NodeJS. NodeJS is not merely a programming language but an asynchronous event-driven JavaScript runtime designed to build scalable applications. In the context of our paper, NodeJS is pivotal for developing and managing the backend services of our digital platform. Its non-blocking I/O model is optimal for handling data-intensive operations, especially when dealing with vast amounts of educational material. Furthermore, its rich package ecosystem, npm, offers various tools and libraries that aid in streamlining our processes, particularly in data reprocessing, which we delve into in sub-Sect 3.

MongoDB. MongoDB serves as our primary database solution, designed to store both the educational content and its associated metadata. As a NoSQL database, MongoDB offers flexibility in storing varied content formats like PDFs, text files, DOCs, among others. This flexibility ensures that our digital library remains versatile and adaptable to different content types. Additionally, MongoDB's document-oriented structure ensures efficient retrieval of content and metadata, essential for real-time recommendation systems.

Data Preparation Process for Digital Library. A robust data preparation process is indispensable for harnessing the true potential of the digital library. This process involves:

1. **Chroma:** Chroma is an advanced tool designed to extract and process key information from educational materials. It focuses on understanding the content's context, structure, and nuances, ensuring that essential knowledge snippets are adequately captured and stored.
2. **Langchain and Chat GPT:** In the realm of Generative AI, Langchain stands out as a sophisticated tool that interlinks the various chunks of knowledge, creating a seamless knowledge chain. Paired with Chat GPT, a state-of-the-art language model, the duo ensures that the digital library remains dynamic, context-aware, and responsive to user queries and needs.
3. **Metadata:** Beyond the core content, the metadata associated with each document is invaluable. Metadata encompasses information like content type, authorship, publication date, and more. This aids in the preprocessing phase, ensuring content is categorized, indexed, and optimized for quick retrieval and recommendation.

Digital Library. The Digital Library, built upon the robust foundation of the aforementioned components, serves as the repository and engine for the E-learning system. Its primary role is to manage, categorize, and retrieve educational material efficiently. Additionally, it prepares prompts and context for the E-learning system, ensuring that when learners interact with the platform, they receive content that is not only relevant but also contextually appropriate.

E-learning System. The E-learning system, as the frontend of our platform, offers a plethora of services tailored for educational training and practice. Leveraging the power of the Digital Library and Generative AI, it provides personalized recommendations, interactive learning modules, and real-time feedback. The system recognizes the unique learning trajectories of individual users, ensuring they receive content that aligns with their current understanding, interests, and future learning goals. Moreover, it facilitates collaborative learning, harnesses community-driven insights, and continually adapts based on user feedback, making the learning experience more immersive and effective.

4   Evaluation Scenarios

In the evaluation of our proof-of-concept, we primarily adopt a quantitative analysis methodology. The raison d'être behind such an analytical choice is twofold: firstly, quantitative methods offer measurable and statistically interpretable outcomes, and secondly, given the nascent stage of our project, it's quintessential to establish a foundational assessment that can be objectively scrutinized.

4.1   Participant Demographics

Our evaluation cohort was composed of a total of 20 students from FPT University's Software Engineering major. A closer inspection of the participant demographics revealed a gender distribution of 16 males (80%) and 4 females (20%). This gender distribution, while not entirely balanced, is reflective of the gender dynamics often observed in many engineering disciplines.

Crucially, the majority of our participants were senior students. This is a salient detail because senior students, with their accumulated academic experience and exposure, are better positioned to provide a nuanced evaluation. Their familiarity with diverse e-learning systems and academic content over the years equips them with a richer context against which to assess our Generative AI-Driven Digital Assistance.

As for the subject matter, our evaluation revolved around a couple of predefined topics extracted from corresponding articles: "Security and Privacy" and "Economy". We took careful measures to ensure that none of the participants had extensive expertise or significant prior experience in these topics. This decision was deliberate, aiming to mimic a genuine e-learning scenario where learners are often introduced to unfamiliar or moderately familiar content. By selecting participants with limited prior knowledge in the selected topics, we aimed to ascertain the effectiveness and relevance of our system's recommendations in a more authentic learning environment.

4.2   Metric of Interest: Recommendation Accuracy

Given the scope and architecture of our system, one might argue for a multi-faceted evaluation encompassing aspects such as response time, scalability, and more. However, we leverage the ChatGPT 3.5 Turbo version, a robust platform that inherently ensures swift response times and high scalability. With these factors being inherently managed by the platform, our evaluation can afford to fixate on the most pivotal metric for any recommendation system: Recommendation Accuracy.

Recommendation Accuracy delves into understanding how congruent the system's suggestions are to the user's actual preferences and needs. For our evaluation, this would manifest in the form of the system's ability to discern and suggest relevant educational content to the students. Precision@k, Recall@k, and F1-Score are typical metrics used to quantify this aspect. In simple terms:

- **Precision@k**: Of the top 'k' recommendations made by our system, how many were deemed relevant by the students?
- **Recall@k**: Of all the relevant recommendations, how many were successfully identified and suggested by our system among its top 'k' recommendations?
- **F1-Score**: A balanced measure, the harmonic mean of Precision and Recall, providing an aggregate sense of the system's accuracy.

By honing in on Recommendation Accuracy, our evaluation aims to ascertain the effectiveness of our system in its primary task, ensuring that the generative AI truly augments the e-learning experience by providing relevant, context-aware suggestions.

4.3 Evaluation Methodology

The evaluation was meticulously designed to ensure comprehensive assessment of our Generative AI-Driven Digital Assistance platform. We split the evaluation into two distinct phases, each tailored to gauge different facets of the system's capabilities.

Phase I: Direct Questions from the Article. The primary phase spanned a duration of 2 h, during which participants engaged with the system using a set of pre-defined questions. These questions, amounting to a total of 50, were crafted in such a manner that they directly referenced content mentioned within the selected articles. The aim of this phase was twofold:

1. To gauge the accuracy and relevance of the system's recommendations when confronted with direct, article-related queries.
2. To ascertain the ease of use and intuitiveness of our platform, given that participants were navigating and interacting with it for the first time.

Given the direct nature of these questions, this phase served as a more controlled environment, allowing us to monitor the system's proficiency in delivering precise and contextually accurate recommendations.

Phase II: Open-Ended Exploration. Transitioning from the controlled environment of the first phase, the second phase offered participants a more open-ended interaction platform. Lasting an hour, this segment encouraged participants to pose random questions pertaining to the broader domains of 'Computer Science' and 'Economy'. Unlike the first phase, here, the questions were not restricted to the content of the articles, providing a more challenging test bed for our AI-driven platform.

The objectives of this phase were:

1. To assess the breadth and depth of the system's knowledge base and its capability to generate relevant recommendations beyond the confines of the selected articles.

2. To evaluate the system's adaptability in addressing a diverse array of questions, gauging its versatility in an uncontrolled setting.

Together, both phases offered a holistic evaluation framework, covering both controlled and open-ended interaction scenarios, thereby providing a thorough examination of the platform's recommendation prowess.

4.4 Results

Table 1. Evaluation Results for the two phases (with $k = 3$)

Student	Phase I			Phase II		
	Precision@3	Recall@3	F1-Score	Precision@3	Recall@3	F1-Score
Student 1	0.85	0.80	0.82	0.80	0.75	0.77
Student 2	0.89	0.85	0.87	0.82	0.79	0.81
Student 3	0.88	0.82	0.85	0.79	0.77	0.78
Student 4	0.90	0.87	0.88	0.85	0.81	0.83
Student 5	0.86	0.80	0.83	0.83	0.78	0.80
Student 6	0.84	0.77	0.80	0.78	0.76	0.77
Student 7	0.87	0.85	0.86	0.80	0.77	0.79
Student 8	0.85	0.82	0.83	0.84	0.80	0.82
Student 9	0.88	0.84	0.86	0.83	0.78	0.81
Student 10	0.89	0.83	0.86	0.82	0.79	0.80
Student 11	0.86	0.80	0.83	0.85	0.82	0.84
Student 12	0.84	0.78	0.81	0.80	0.77	0.78
Student 13	0.83	0.79	0.81	0.81	0.75	0.78
Student 14	0.85	0.81	0.83	0.82	0.76	0.79
Student 15	0.87	0.83	0.85	0.83	0.80	0.81
Student 16	0.88	0.82	0.85	0.80	0.78	0.79
Student 17	0.84	0.79	0.81	0.79	0.74	0.76
Student 18	0.86	0.80	0.83	0.84	0.79	0.82
Student 19	0.85	0.81	0.83	0.81	0.77	0.79
Student 20	0.87	0.82	0.84	0.83	0.79	0.81

From the evaluation table (Table 1), we discern several pivotal observations regarding student performance over the two phases. In Phase I, Student 4 stood out with the highest F1-Score of 0.88, underpinned by a notable Precision@3 of 0.90. Contrarily, Student 6 displayed the lowest efficiency in this phase with an F1-Score of 0.80. Transitioning to Phase II, Student 11 led the pack, achieving an F1-Score of 0.84, while Student 17 lagged with the lowest score of 0.76. A cursory glance reveals that students generally fared better in Phase I, presumably due to the direct article-related questions, as opposed to the broader, more challenging queries in Phase II. Additionally, a recurring trend across both phases was the system's higher precision scores compared to recall, implying that while the top 'k' recommendations were largely accurate, there's potential for augmenting the recall for a more comprehensive recommendation set.

5 Conclusion

Our exploration into Generative AI mechanisms, building on the foundational strides of prior research, emphasizes the potency of combining traditional recommendation algorithms with cutting-edge generative techniques. The potential of our Generative AI-driven methodologies, as showcased in this manuscript, signifies a promising trajectory for E-Learning digital assistance. The meticulous evaluation with FPT University students substantiates the potential of this synthesis to redefine digital learning. In culmination, we stand at the precipice of a new E-Learning era, harmonizing classical methodologies with innovative generative techniques to cater adeptly to the modern scholar's multifaceted needs.

References

1. Akbar, M., et al.: Recommendation based on deduced social networks in an educational digital library. In: IEEE/ACM Joint Conference on Digital Libraries, pp. 29–38. IEEE (2014)
2. Almaghrabi, M., Chetty, G.: Deep machine learning digital library recommendation system based on metadata for Arabic and English languages. In: 2020 IEEE Asia-Pacific Conference on Computer Science and Data Engineering (CSDE), pp. 1–6. IEEE (2020)
3. Bulut, B., et al.: User's research interests based paper recommendation system: a deep learning approach. In: Putting Social Media and Networking Data in Practice for Education, Planning, Prediction and Recommendation, pp. 117–130 (2020)
4. Heylighen, F., Bollen, J.: Hebbian algorithms for a digital library recommendation system. In: Proceedings of the International Conference on Parallel Processing Workshop, pp. 439–446. IEEE (2002)
5. Hu, J., et al.: Research on reading subject recommendation of library based on data analysis. J. Phys. Conf. Seri. **1607**, 012022 (2020)
6. Huang, Z., et al.: A graph-based recommender system for digital library. In: Proceedings of the 2nd ACM/IEEE-CS Joint Conference on Digital Libraries, pp. 65–73 (2002)
7. Jomsri, P.: Book recommendation system for digital library based on user profiles by using association rule. In: Fourth edition of the International Conference on the Innovative Computing Technology (INTECH 2014), pp. 130–134. IEEE (2014)
8. Lee, J., et al.: Personalized academic research paper recommendation system. arXiv preprint arXiv:1304.5457 (2013)
9. Liu, M.: Personalized recommendation system design for library resources through deep belief networks. Mob. Inf. Syst. **2022** (2022)

10. Pham, T., et al.: Digital transformation in engineering education: exploring the potential of AI-assisted learning. Australas. J. Educ. Technol. **39**(5), 1–19 (2023)
11. Tejeda-Lorente, Á., et al.: A quality based recommender system to disseminate information in a university digital library. Inf. Sci. **261**, 52–69 (2014)
12. Thanh, B.N., et al.: Race with the machines: assessing the capability of generative AI in solving authentic assessments. Australas. J. Educ. Technol. **39**(5), 59–81 (2023)
13. Wang, J., et al.: A systematic review of recommendation system based on deep learning methods. In: Nedjah, N., Martínez Pérez, G., Gupta, B.B. (eds.) ICSPN 2022. LNNS, vol. 599, pp. 122–133. Springer, Cham (2023). https://doi.org/10.1007/978-3-031-22018-0_12
14. Wayesa, F., et al.: Pattern-based hybrid book recommendation system using semantic relationships. Sci. Rep. **13**(1), 3693 (2023)
15. Zhao, L.: Personalized recommendation by using fused user preference to construct smart library. Internet Technol. Lett. **4**(3), e273 (2021)

Digital Transformation of Teaching and Learning in Higher Education Institutions: A Case Study of the University of Naples "Parthenope"

Danilo Greco[1], Aizhan Tursunbayeva[2], Rosita Capurro[2],
and Maricarla Staffa[3]

[1] Department of Management, Economics and Industrial Engineering, Politecnico di Milano, Via Lambruschini 24/b, 20156 Milan, Italy
danilo.greco@polimi.it
[2] DISAE - Department of Business and Economic Studies - Università degli Studi di Napoli Parthenope, Via Generale Parisi, 13, Napoli, Italy
{a.tursunbayeva,rosita.capurro}@uniparthenope.it
[3] DiST - Department of Sciences and Technology - Università degli Studi di Napoli Parthenope, Centro Direzionale, isola C4, Napoli, Italy
mariacarla.staffa@uniparthenope.it

Abstract. This paper investigates the digital transformation journey of the University of Naples Parthenope in Italy through an in-depth qualitative case study. The analysis adopts a holistic socio-technical perspective encompassing strategic, cultural, structural, human resource, and technological dimensions across the key activities of teaching, research, and operations. The study addresses a gap in current knowledge by developing an integrative digital transformation framework for universities grounded in the literature. Data collected from questionnaires about the adoption of Digital Technologies as well as innovative teaching formats revealed multi-level barriers and enablers influencing the transformation process. Key findings indicate that developing new digitally enabled models of teaching enhanced productivity and competitiveness. However, realizing the full potential of emerging technologies requires forward-thinking leadership, organizational agility, new capabilities, and a supportive culture. This study contributes original empirical insights and an actionable framework for leveraging technologies strategically to enhance the competitiveness and sustainability of universities during a period of rapid digital change. This pilot study will be extended in the future by delving into the more generic role of Artificial Intelligence (AI) in the university process.

Keywords: Digital transformation · Strategic and Organizational changes · Digital Technologies · Human resource management · Innovation · University

F. Palomba and C. Gravino (Eds.): WAILS 2024, LNCS 14545, pp. 99–111, 2024.
https://doi.org/10.1007/978-3-031-57402-3_11

1   Introduction

Artificial Intelligence has emerged as a transformative force in various sectors, and the educational landscape is no exception. Integrating AI into university processes can significantly enhance efficiency, decision-making, and overall student experience. For instance, AI-powered learning management systems can provide personalized educational content, adaptive assessments, and timely feedback, catering to individual student needs. Additionally, AI algorithms can streamline administrative tasks, optimizing resource allocation and improving overall institutional effectiveness. For this reason, higher education institutions worldwide face mounting pressures to innovate and transform amidst rapid technological changes by disrupting conventional university models and practices [8]. However, prior research suggests many universities are slow to harness these digital technologies strategically to enhance competitiveness, productivity, and sustainability [4,9]. This inability to adapt poses risks in an increasingly dynamic landscape.

While studies have examined discrete applications of digital technologies in universities, few take a holistic approach to investigating the organizational implications [7]. This paper addresses this gap through an in-depth case study of the digital transformation journey of the University of Naples Parthenope (from now on *Parthenope*) in Italy.

The overarching research question is: how can Parthenope strategically harness digital technologies to drive organizational innovation and change for enhanced competitiveness and sustainability?

This study is part of a larger project titled *Towards a Digital, Sustainable, Intelligent, and Inclusive University: Strategic, Organizational, and Technological Intersections for Competitiveness and Success*, whose intent is to answer the following main questions:

- What are the barriers and enablers at organizational, team, and individual levels influencing the digital transformation process?
- How can digital capabilities be developed and integrated across teaching, research, and operational activities?
- What human resource management practices and changes facilitate adopting new technologies and work practices?
- What strategies, leadership approaches, and change management initiatives are most effective in guiding transformation?
- How does digitalization impact organizational culture, structure, and business models?

As a first step, in this work, we present a mixed-method research study in which we measure the tendency of Parthenope to adopt digital technologies or innovative teaching methods and we analyzed the perception of both students and teachers to better understand whether a promotion of these technologies can have a positive impact on Parthenope's organizational processes. We leave consideration about the role of Artificial intelligence in future work.

This study provides first original theoretical contributions pavyng the way for the developing an integrative digital transformation framework for universities encompassing technological, human, and organizational factors. The presented empirical investigation of Parthenope delivers practical insights into how higher education institutions can holistically leverage technologies to their strategic benefit.

The paper is structured as follows. First, a critical review of relevant literature on the digital transformation of universities is provided. The case study methodology is then described, followed by the presentation of the key results. Finally, the theoretical and practical implications are discussed along with limitations and future research needs.

2   Literature Review

The higher education landscape is undergoing rapid change spurred by emerging digital technologies. Recent research indicates that capabilities such as data analytics, automation, artificial intelligence, extended reality, and the Internet of Things are transforming learning and teaching models, research practices, and university operations [4,6,9]. However, surveys reveal many universities lag in digitally enabling these core activities to enhance productivity, sustainability, and competitiveness [4,7]. From a strategic perspective, digital transformation is highlighted as an organizational imperative for innovation, growth, and survival amidst exponential technological changes [10]. Yet competing value priorities between academic excellence versus efficiency gains often create tensions [2]. Studies emphasize the need for ambidextrous strategies balancing exploration and exploitation [1]. Leadership is also critical in providing direction while allowing autonomy [9]. Structurally, digitalization necessitates greater collaboration, new governance models, and organizational agility [8]. Digitally mature universities exhibit decentralized decision making, interdisciplinary programs, flexible structures, and administrative data integration [9]. Developing a digital culture encourages experimentation and continuous learning is vital for change [1]. From a human resource perspective, digital literacy across all staff is foundational. Researchers stress that technology alone is insufficient; successfully leveraging emerging technologies requires investment in human capabilities [8]. Fostering digital skills and mindsets enables innovation and reduces resistance [10]. In summary, current literature recognizes digital transformation in higher education as a complex socio-technical challenge requiring coordinated development of technologies, individuals, and organizational environments. This study addresses knowledge gaps by taking a holistic perspective and empirically examining the interplay of strategic, cultural, structural, human, and technological factors influencing the digital journey.

3   Methodology

This study adopts a single-case study design to empirically investigate the digital transformation at the University of Naples Parthenope. This approach is fitting for examining technology-driven organizational change processes, capturing perspectives from different actors, and providing rich insights into a complex phenomenon in context [11]. Data is planned to be collected for this case study from multiple sources. An online mixed-method survey was designed and administered to a sample of 151 students, 20 academics to gauge technology readiness, awareness, attitudes, skills, and training needs. The survey was written in Italian and was designed based on the existing established questionnaires piloted in other contexts. Descriptive statistics were calculated for quantitative measures, while qualitative data was thematically analyzed. This survey is the first step of a large-scale research project funded by the University of Naples Parthenope. The follow-up research phases will involve analysis of the relevant documents, reports, and internal communications were also reviewed to provide background and verify findings. Ethical protocols around confidentiality, anonymity and informed consent have been considered to ensure that our research meets the recommendations for conducting research involving human participants and emerging digital technologies such as AI. In summary, such a comprehensive and systematic methodological approach will permit to development of rich, contextualized insights into Parthenope's digital transformation journey from both managerial and staff perspectives, while gathering data from multiple sources to enable robust analysis and interpretation.

4   Findings

4.1   Analyzing Student Performance Data from an Innovative Teaching Tools University Study

This section presents an analysis of the survey responses from 151 university italian students regarding their use and perceptions of innovative teaching methodologies and digital technologies. The survey questions pertain to the students' year and field of study, problems encountered with traditional lecturing, participation in innovative teaching activities, opinions on such activities, and preferences for future adoption of digital technologies.

Most respondents hailed from undergraduate programs in Business Administration, Computer Science, and Management. The primary issues cited with traditional pedagogy were a lack of student engagement, incentives for participation, and practical activities (Fig. 1).

Contrary to what we expected, a fair percentage (39.1%) of the students pointed out that innovative teaching methods had been adopted (Fig. 2).

Commonly implemented innovative activities included business simulations, case studies, and team-based assignments (Fig. 3).

As a major outcome of this study, students broadly praised these innovations as enhancing engagement and providing practical applied learning (Fig. 4). About 64% of the students advocated increased integration of both innovative teaching and digital technologies to further improve the educational experience.

Suggested technologies included e-learning platforms, videoconferencing, multimedia resources, and online collaboration tools. Overall, the data indicates

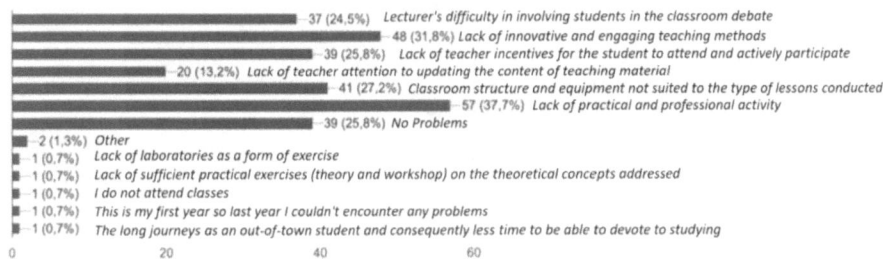

Fig. 1. Histogram of problems encountered by students with traditional lecturing

students' receptiveness to innovative teaching approaches leveraging technology for heightened interactivity, motivation, and real-world skills development. Further qualitative and longitudinal study is warranted to assess the efficacy of specific methodologies on student performance and satisfaction across educational contexts.

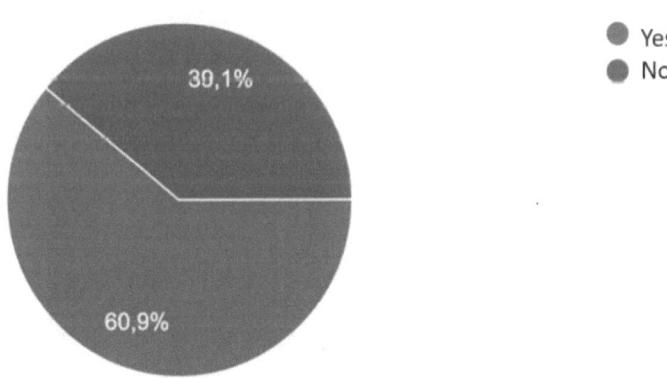

Fig. 2. Pie Chart of innovative teaching adoption

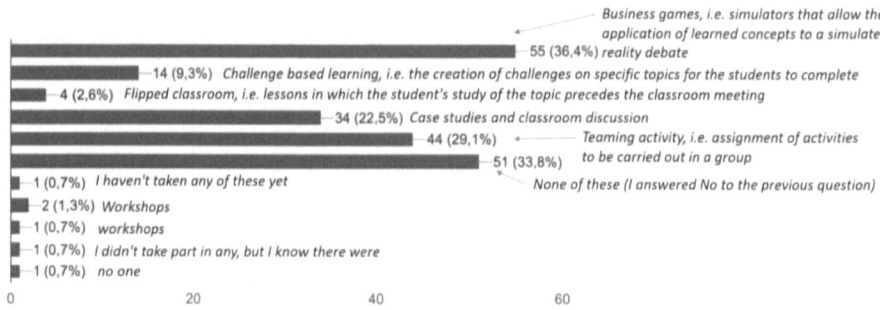

Fig. 3. Innovative teaching activities

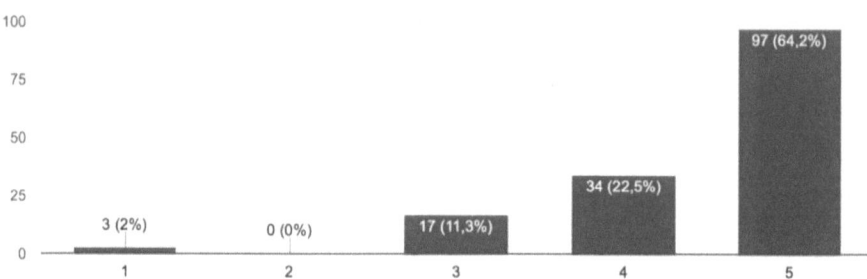

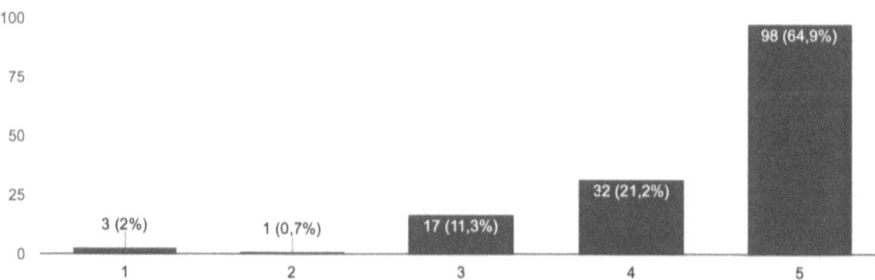

Fig. 4. Willingness to adopt up) innovative Teaching and down) Digital Technologies in Teaching contexts.

In summary, the figures showed results from a survey of university students regarding their use and perceptions of innovative teaching tools. From the survey, it emerged the usage of various tools by year of study such as chatbots, virtual reality, and interactive online materials. Additionally, survey results demonstrate that students gave the highest ratings for the effectiveness of innovative teaching tools, with over 60% saying they are extremely effective. Overall, the data indicates that students find innovative teaching tools more effective than first-year students.

4.2   Analyzing Teachers' Use of Innovative Didactic Tools in the University

We here present the discussion of the survey responses from 20 university instructors regarding their adoption of innovative teaching methodologies and digital technologies. Respondents hailed predominantly from fields like computer science, physics, and business administration. Commonly cited challenges with traditional pedagogy included poor class attendance and student disengagement. The most frequently implemented innovative techniques were case studies, team-based assignments, and business simulations (Fig. 5).

With reference to last year, indicate the problems you encountered with the classroom experience (maximum 3 answers)

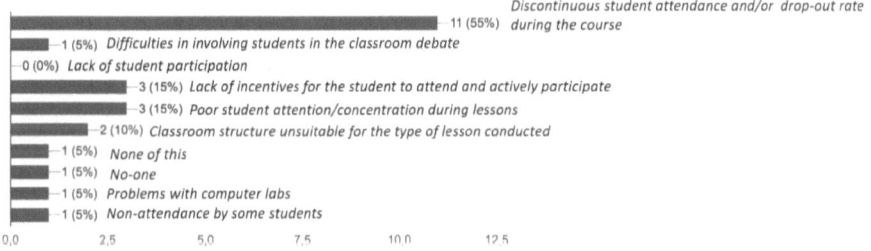

Fig. 5. Histogram of problems encountered by Professors in regards to students.

Surprisingly, the 85% of Teachers adopted innovative Teaching methodologies (Fig. 6) such as Case Study and team activities (Fig. 7).

Among them, a great part (90%) reported integrating digital technologies such as e-learning platforms, videoconferencing, and multimedia resources to enable innovative instruction (Fig. 8).

Looking ahead, the majority of instructors plan to employ both innovative methods and supporting digital technologies in their future courses. However, few had implemented extracurricular activities to engage students further.

Summarising, the data reveals a broadly favourable attitude among university faculty towards pedagogical innovation and technology integration to enhance learning experiences. Considerable potential remains for advancing the implementation of student-centred, digitally-enabled active learning techniques within higher education. Further research should explore specific best practices

Indicate whether you adopted innovative teaching tools during
last year's lessons. 20 answers

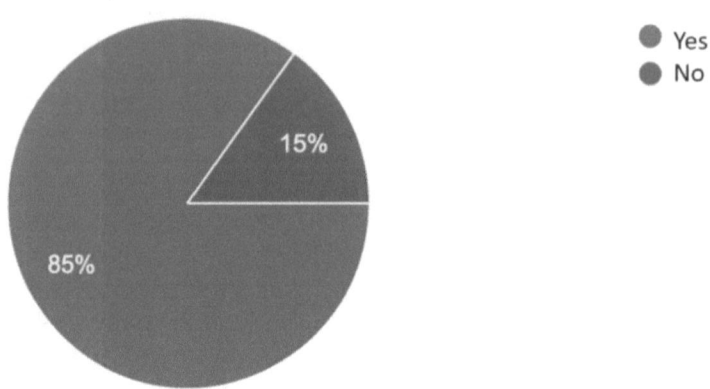

Fig. 6. Pie Chart showing the percentage of Professors adopting innovative teaching
methods.

Select the innovative teaching methods you adopted last year. 17 answers

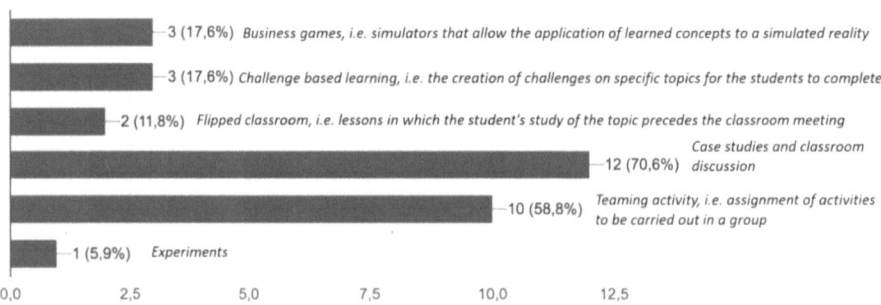

Fig. 7. Innovative didactic techniques.

for the sustainable adoption of instructional innovations across institutional contexts. Popular innovative methods implemented included case studies, team-
based assignments, and simulations. Approximately half of the respondents had
incorporated technologies like e-learning platforms, videoconferencing, and multimedia resources. Looking ahead, most instructors intend to employ both innovative techniques and supporting digital technologies in future courses. However,
few had instituted extracurricular initiatives to engage students further. Overall, the data reveals a generally favourable attitude among university faculty
towards pedagogical innovation and technology integration to enhance learning. Considerable potential remains for expanding the implementation of active,
student-centred learning approaches facilitated by digital tools in higher educa-

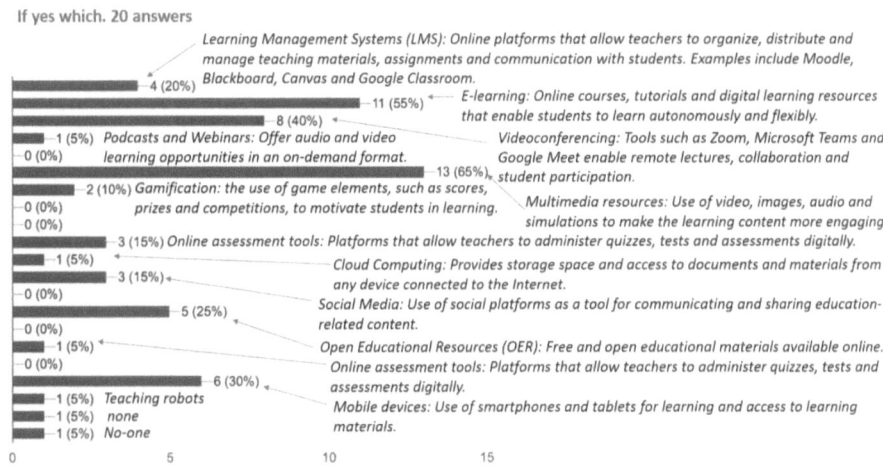

Fig. 8. Adopted Digital Technologies.

tion contexts. Further research should investigate specific best practices for the sustainable adoption of instructional innovations across institutions.

4.3   Considerations on the Use of DT Within Parthenope's University

The analysis of the questionnaires revealed several key themes related to Parthenope University's digital transformation:

Strategic Approach

- A centralized e-learning strategy increased online course offerings, enabled remote learning, and improved student satisfaction. However, technological integration in classroom teaching was slower, indicating a need for further training and incentives for academics.
- Research teams that proactively experimented with digital tools like data analytics, modelling, and visualization achieved greater productivity and success with publications and grants.
- Legacy administrative processes underwent automation and digitization, yielding cost and time efficiencies. But back-office systems remained siloed, reducing institutional agility.

Organizational Culture

- Students and younger academics emerged as champions of technology adoption, exerting bottom-up pressure for innovation.
- Resistance from some senior faculty and administrators stemmed from perceived loss of status, scepticism about efficiency gains, and wariness of transparency.

– Greater openness to experimentation, collaboration, and multi-disciplinary programs was evident in departments where leaders role-modelled digital behaviours.

Structure and Governance

– Centralized IT services successfully implemented institution-wide platforms like the learning management system and research portal. However, the limited involvement of end-users led to integration challenges.
– Grassroots innovation often occurred in departments with decentralized resources and autonomy, but best practices were not scaled due to siloed structures.
– Governance bottlenecks were caused by traditional committee structures that lagged in technology advancements.

Human Resources

– Junior researchers and lecturers developed greater digital literacy through informal peer learning, but mid-career faculty's skills stagnated without training incentives.
– Administrative staff struggled with new digital systems that altered workflows without adequate change management support.
– Growing demand for specialized skills like data science, AI development, and learning designers necessitated targeted recruitment and development.

5   Discussions

This study provides important insights into the multifaceted process of digital transformation in universities. The key findings and implications are discussed below:

Theoretical Implications

– The analysis confirms prior research emphasizing digital transformation's sociotechnical nature spanning technology, individuals, and organizations [8]. The proposed framework integrates these elements to fill a knowledge gap.
– The change process was gradual and met resistance, aligning with theories on technology acceptance and diffusion of innovations [5]. Both bottom-up and top-down strategies drove adoption.
– Ambidextrous practices were evident, as suggested by studies on organizational agility during technological change [1]. Exploration and exploitation required balancing.

Practical Implications

– Universities should audit digital skills across academic fields and functions to target training where needed most to avoid imbalances.
– Leadership development is vital to building digitally-savvy change leaders able to communicate a compelling vision.
– Governance models need realignment with digital priorities through cross-functional teams and reduced bureaucracy.
– Investment in specialized talent like data scientists and AI experts is essential to capitalize on emerging technologies.
– Holistic digital strategies are needed to break down structural and cultural silos and enable the scalability of innovations.

In summary, supporting university digital transformation requires coordinated development of new technology capabilities, individuals' digital skills, and digitally mature organizational cultures, structures, and leadership.

Limitations and Future Research. The presented survey is only the first step of a large-scale research project funded by the University of Naples Parthenope. The follow-up research phases will involve analysis of the relevant documents, reports, and internal communications were also reviewed to provide background and verify findings. Specifically, we plan to conduct a multi-level analysis to examine the digital transformation issues at the organizational, departmental, work unit, and individual job role levels. The qualitative data to be collected will be coded using NVivo and analyzed iteratively following established principles of theme identification, comparison, and interpretation [3]. We also plan to conduct semi-structured interviews with a sample of 15 individuals selected through purposive sampling, including university leaders, heads of departments, academics and administrative staff. Questions here will elicit perspectives on digital transformation strategies, barriers, enablers, and impacts and will incorporate consideration about the adoption of AI at large. Infatc, another limitation of this work is that, while the study provides valuable insights, a more explicit connection to the role of AI is necessary for outlining the potential applications of AI in the context of the university processes studied. This could provide some insights on how AI solutions might address the limitations identified in the current processes and improve outcomes. Also, expanding the participant pool to encompass a more diverse range of stakeholders is crucial when discussing the implications of AI in the university setting. In doing so, the study could capture varied perspectives, ensuring a more holistic understanding of the challenges and opportunities presented by AI. Diverse participants could include students, faculty, administrative staff, but also AI experts, each can offer unique insights into the subject matter. In future studies, we aim at involving at least 5 focus groups of 6–8 participants composed of the aforementioned people categories to explore viewpoints on technology and AI adoption. The sessions will be audio recorded and transcribed. Such mixed-method and multi-phased research design will permit triangulate across different data sources and methodological traditions increasing the validity and credibility of the research findings. We will bolster the study's robustness, providing demographic statistics of the participants is paramount. This not only adds transparency to the research process but also allows for a nuanced analysis of how different demographic groups perceive and interact with AI-driven changes in the university processes. Understanding the nuances among demographics can be pivotal in tailoring AI implementations to be more inclusive and equitable.

Another limitation is that this study focused on a single university case. Additional research should test the digital transformation framework in other higher education contexts. Longitudinal designs could also assess the change process over time. Comparative analyses of universities at different stages of digital maturity may offer useful insights.

6   Conclusions

This study investigated the digital transformation journey of the University of Naples Parthenope through an in-depth qualitative case study. A key initial finding was that leveraging emerging digital technologies to enhance competitiveness and sustainability requires coordinated development of technological capabilities, individuals skills, and organizational change. While Parthenope made gains in digitally enabled teaching models, research tools, and streamlined processes, multiple barriers spanning strategy, culture, structure, and human resources were also uncovered. This highlights the complex, multifaceted nature of digital transformation. The proposed integrative framework provides a model for assessing and guiding digital maturity across the key dimensions of technology adoption, strategic alignment, organizational agility, workforce capability, and cultural readiness. Focusing holistically on these interdependent elements is critical for university leaders seeking to harness the transformative power of digital technologies. The study's contributions include new empirical insights into the digital transformation challenges and opportunities facing higher education institutions and an actionable framework. However, the research is limited to a single case study. Further investigation should test the framework in other university contexts.

In conclusion, rapidly emerging technologies present risks but also enormous potential for universities to enhance productivity, sustainability, global reputation, and competitiveness. This study provides a roadmap for fully leveraging digital capabilities to drive innovation. A coordinated, holistic approach is key to realizing the promise of transformation.

In future studies, we aim at strengthening the narrative around the impact of AI on university processes, refining survey and figure presentation, and embracing a more diverse participant pool with demographic statistics, with the aim of evolving the study into a comprehensive exploration of the intersection between traditional educational processes and the transformative potential of Artificial Intelligence.

Acknowledgements. This work was supported by the research project "Towards a Digital, Sustainable, Intelligent, and Inclusive University: Strategic, Organizational, and Technological Intersections for Competitiveness and Success," which was funded by the University of Naples Parthenope under research funding for young researchers' career development.

References

1. Borzillo, S., Schmitt, A., Antino, M.: Communities of practice: keeping the company agile. J. Bus. Strategy **33** (2012). https://doi.org/10.1108/02756661211281480

2. Tømte, C.E., Trine Fossland, P.O.A., Degn, L.: Digitalisation in higher education: mapping institutional approaches for teaching and learning. Qual. Higher Educ. **25**(1), 98–114 (2019). https://doi.org/10.1080/13538322.2019.1603611

3. Miles, M., Huberman, A., Saldana, J.: Qualitative Data Analysis. SAGE Publications (2014). https://books.google.it/books?id=3CNrUbTu6CsC

4. Nwankpa, J.K., Roumani, Y.: It capability and digital transformation: a firm performance perspective. In: International Conference on Interaction Sciences (2016). https://api.semanticscholar.org/CorpusID:2200077

5. Rogers, E.M., Singhal, A., Quinlan, M.M.: Diffusion of innovations. In: An integrated Approach to Communication Theory and Research, pp. 432–448. Routledge (2014)

6. Rossi, S., Capasso, R., Acampora, G., Staffa, M.: A multimodal deep learning network for group activity recognition. In: 2018 International Joint Conference on Neural Networks (IJCNN), pp. 1–6 (2018). https://api.semanticscholar.org/CorpusID:52984570

7. Selwyn, N.: Should Robots Replace Teachers?: AI and the Future of Education, 1st edn.. Polity Press (2019)

8. Stuart, E., Phillips, T., David, R.: How data and digital technologies can transform education systems. In: Ra, S., Jagannathan, S., Maclean, R. (eds.) Powering a Learning Society During an Age of Disruption. EARICP, vol. 58, pp. 311–321. Springer, Singapore (2021). https://doi.org/10.1007/978-981-16-0983-1_21

9. Tungpantong, C., Nilsook, P., Wannapiroon, P.: A conceptual framework of factors for information systems success to digital transformation in higher education institutions. In: 2021 9th International Conference on Information and Education Technology (ICIET), pp. 57–62 (2021)https://doi.org/10.1109/ICIET51873.2021.9419596

10. Vial, G.: Understanding digital transformation: a review and a research agenda. J. Strat. Inf. Syst. **28**(2), 118–144 (2019). https://doi.org/10.1016/j.jsis.2019.01.003, https://www.sciencedirect.com/science/article/pii/S0963868717302196, sI: Review issue

11. Yin, R.K., Campbell, D.T.D.T.: Case Study Research and Applications: Design and Methods. SAGE Publications, Inc., 6th edn. Thousand Oaks, California (2018)

EDUSOCIAL & IA The Educational Use of Artificial Intelligence in Social Media

Alfonso Amendola(✉) (iD)

Department of Business Sciences, Management and Innovation System, University of Salerno, Salerno, Italy
alfamendola@unisa.it

Abstract. The advent and development of technology have revolutionized the means and techniques for disseminating information. Social media have established themselves, becoming the main scenarios for debate, sources of dissemination of new ideas and virtual places through which individuals exchange views and opinions sitting from their armchairs at home. However, since they are platforms accessible to all, the difficulties in managing content considered illegal or violent have been amplified. The essay aims to analyze the theme of Social Media through the universe of Artificial Intelligence in its educational perspective. Highlighting risks and prospects.

1 A Theoretical Premise

Educational use of social media as an opportunity has long been reflected upon, starting with personalization of learning and access to relevant content [1, 3, 5, 7–10]. Developments in artificial intelligence have made possible further implementation of new forms of learning and interaction on social media. These platforms can offer a range of educational benefits through the use of artificial intelligence in different ways and practices. One of the main functions of artificial intelligence in social media is the personalization of the learning experience. By analyzing users' interests and behaviors, artificial intelligence can provide targeted and relevant educational content. This allows users to access educational material specific to their needs and preferences, thus enhancing individualized learning. In addition, artificial intelligence can be used to identify new trends and topics of educational interest based on user data. Giving additional opportunities for educators and students to stay up-to-date with the latest news and events in education.

In addition, artificial intelligence can also suggest reliable and quality sources of information, helping to combat the spread of false or unreliable news. However, it is also important to consider the challenges associated with using artificial intelligence in social media for educational purposes. For example, users' privacy could be compromised when their personal data is collected and analyzed to personalize the educational experience. Therefore, it is critical to ensure data protection and respect for users' privacy when using artificial intelligence in social platforms. But let's proceed in order.

2   A First Socio-cultural Definition

Although we owe the definition of the concept of Artificial Intelligence to Turing [6], the term was coined by John McCarthy, who was the first to recognize AI as a new and specific field of study. Summer of 1951: "the science and engineering of creating intelligent machines". Its definition remains over time. And it indicates that component of Computer Science that deals with simulating intelligent behavior in computers, specifically with programming hardware and software systems that provide machines with typically human characteristics, such as language capabilities, visual, space-time and decision-making perceptions. It can also be defined as the set of machine skills such as learning, reasoning, planning and creativity developed through observing the effects of previous actions and working autonomously. Specifically, the English Oxford Dictionary defines it like this: «The theory and development of computer systems able to perform tasks normally requiring human intelligence, such as visual perception, speech recognition, decision-making, and translation between languages». Anyway, McCarthy and other U.S. researchers organized a conference. The purpose of this conference is to discuss key issues related to automation, neural network development, and human intelligence studies. In the paper presented at the conference, the fundamental goal of the project is clear: to find out how to make machines capable of using language and solving typically human problems. Basically, they aimed to understand how to develop a mechanical system that could replicate specific traditionally human functions. We have to wait until 1959 when Arthur Samuel created the first software capable of learning from its own experience, representing one of the first examples of machine learning in the history of Artificial Intelligence. In the same year, A. Newell and H. A. Simon developed the General Problem Solver (G.P.S.), the first computer program designed to emulate the human problem-solving approach. Nevertheless, the real turning point in the history of AI is realized with Perceptrons by Minsky and Papert (1969). Here it is shown that despite the progress made in the field, existing systems were still unable to function like the human mind. This event marks the beginning of the so-called "Winter of Artificial Intelligence." That period highlighted the gap (still persisting today) between advanced theories and the practical application of AI, a gap underestimated by researchers, who had been misled by the promises associated with the development of these technologies. For these reasons, Artificial Intelligence remained in a period of stagnation in the field of computer science for more than a decade, particularly with regard to the development of so-called Strong AI. The idea of creating artificial systems that could fully replica the functioning of the human brain was abandoned in favor of a more pragmatic approach based on analyzing the specific functions that AI could more easily replicate. Today, Artificial Intelligence is undoubtedly considered one of the most important innovations of the 21st century, capable of radically transforming not only the way we access information but also society itself. This extraordinary field of technology has proven to be a key factor in a number of areas, including social media and online content management, as well as becoming increasingly embedded in our daily lives [2, 11, 12]. With regard to social media, AI has assumed a role of crucial importance. Its ability to analyze and interpret an impressive volume of user- generated data is far beyond human capabilities. Through real-time processing of text, images and videos, it can identify content that might violate the platforms' guidelines. This initial oversight allows it to quickly identify explicitly harmful or non-compliant content, limiting its visibility or removing it entirely. What makes AI even more remarkable is its ability to adapt to the ever-changing nature of online information. It is not limited to static evaluation, but embarks on a continuous learning path. By leveraging sophisticated algorithms and machine learning models based on neural networks, it is able to accurately distinguish between acceptable and inappropriate content, improving its ability to recognize new forms of harmful or offensive content over time. However, the decision to deploy this technology has affected so many other areas, such as the automation of self-driving vehicles, the use of autonomous systems in business and finance, the use of facial recognition in the fight against terrorism, and the application of intelligent systems in the field of healthcare, all of which demonstrate that artificial intelligence is no longer something that concerns science fiction, but has become an integral part of our daily lives. It remains important to point out, however, that this omnipresence also has important implications in the political and public context [4].

3   Towards Digital Platforms

Digital platforms, often controlled by a small number of actors, use AI algorithms to filter and personalize political information and messages. This can limit citizens' exposure to diverse opinions and limit their critical sense. The lack of quality and variety of information in public debate, combined with the possible targeted circulation of fake news, can undermine pluralism of opinions and values, affecting the proper functioning of the democratic system. In summary, Artificial Intelligence is a driving force in online content management and represents one of the most significant innovations of our time. This technology, with its capacity for advanced processing and continuous learning, is capable of shaping the way we live, inform ourselves and participate in public debate, addressing complex challenges and contributing significantly to our ever-changing society.

Returning to the discussion of the right to freedom of expression, we can say that it has undergone various modifications over the years, adapting to the changing world of communication and information, modifications due to the fact that times change and with them the dynamics involved. It turns out to be important, therefore, to examine whether and how AI technologies interact with this fundamental principle of democracy. As anticipated, this is a rather complex objective since, to identify specifically the different forms of Artificial Intelligence involved in communication and information processes is very difficult; moreover, the right to freedom of expression encompasses a multiplicity of other rights, which we could define as "minor", including the freedom of opinion, the freedom to communicate and express one's thoughts or the freedom to inform and be informed. All of these, as already largely anticipated, have been subject to transformations over the past decades, mainly due to the development of the media.

4   IA and Communication

To better understand the different roles that AI technologies can play in the world of communication and information, a distinction must be made between two different categories. Indeed, they can produce information or discourse to be communicated, or they can select and sort online content, deciding its circulation and dissemination. Regarding the first role, we can say that it is still under development. A very famous example dates back to September 2020, the month when the "Guardian" decided to publish an article entirely written by an AI-based language generator. In the article, the software specified that AI and robots were not threats, reassuring humans with a speech explaining the coming in peace of these new technologies. Although the system had demonstrated great capabilities in the area of coherence of topics covered, scholars said there was still skepticism about the ability to understand the messages.

Although even today, these technologies are in the early experimental stage and far from being fully assimilated into human language, it is important to note the steady increase of algorithms working to generate results in the form of conversations, e.g., chatbots such as Amazon's Alexa, Apple's Siri, Google's Assistant, and Microsoft's Cortana. Issues related to the discourses produced by these algorithms will most likely grow and become increasingly complex as the forms of expression created by autonomous systems become more "intelligent" and cause implications. The second role, which is also very important, concerns the ability of these technologies, to select and sort expressions, different opinions and information published on the web. In this case, Artificial Intelligence does not generate actual content, but filters it, deciding whether and how it can circulate in the digital world, moving from social networks to search engines. An important function is to select, just on social media, content that is considered relevant and popular or offensive and harmful, the former always highly visible and often sponsored in the main sections, while the latter promptly deleted and unreachable. As anticipated, AI plays a key role in moderating online content, an essential function in the digital age where information sharing is ubiquitous. Initially, AI acts as a virtual defense shield, performing an important supervisory and security function within online platforms. Its primary task is to conduct rapid and accurate scanning through large amounts of text, images, and video, with the ultimate goal of detecting potential violations of policies established by the platform. This scanning, which can cover large volumes of content in a very short amount of time, is critical to ensuring users enjoy a safe virtual environment that complies with current guidelines. However, the role of AI in moderation is not limited to only superficial verification. Its function takes on a dynamic and evolving dimension as it is designed to learn continuously over time.

5 Between "Deep Learning" and "Machine Learning"

This means that, thanks to its advanced processing capabilities, AI does not just identify and eliminate inappropriate or offensive content, but constantly adapts to recognize new manifestations of such content. This adaptation is possible through the application of sophisticated machine learning techniques and the use of neural networks, which enable it to understand the broader context in which content is found and detect its nuances. To better understand this concept, it is necessary to make a small digression that delves into the concepts of "deep learning" and "machine learning". When we talk about the former, we mean to explain the ability of systems based on artificial neural networks to learn and recognize similar traits that exist between elements of an inhomogeneous set and to divide them into more homogeneous subsets. This technique thus enables a system to analyze large amounts of data, classify them according to the common elements that characterize them and thanks to this formulate possible predictable solutions.

As for "machine learning," on the other hand, it identifies the set of techniques that enable artificial systems to acquire information directly from given examples, from the data provided to them and from its experience. In this way, the system is able to learn new features intelligently, that is, without the need to follow predetermined rules. The key factor concerning the development of these techniques lies in enabling the creation of such intelligent systems that can collect data from the outside world and, after analyzing its contents, make autonomous decisions. Indeed, Artificial Intelligence can recognize the evolution of linguistic expressions and visual content formats, ensuring that moderation is effective even in an ever-changing digital landscape. This ability to constantly adapt and improve is critical to meeting the constantly changing challenges of the online world. In particular, AI can detect changes in language, detecting emerging expressions or forms of communication that might violate platform policies. In addition, it can analyze large amounts of data to detect growing trends and forms of potentially harmful content that might escape the human eye. All of this helps ensure that online content, capable of undermining the safety or experience of users, is detected and dealt with accurately and promptly. AI acts as a sentinel, ensuring that the virtual community can navigate in a safe online environment that conforms to established guidelines. Its ability to constantly learn and adapt is critical to maintaining the relevance and effectiveness of online content moderation, ensuring that users can enjoy a positive and risk-free digital experience. In the society we live in, the sharing of digital content is an integral part of daily life, and the implementation of AI in content moderation is very helpful in achieving a state of virtual balance.

6 Digital Age

In the digital age in which we live, new technologies have revolutionized the way communication and information are disseminated and perceived. The emergence of the "post-truth" era has radically transformed the media landscape, challenging the traditional rules that once presided over conventional news media. The concept of truth, based on objective and verifiable facts, seems to have been replaced by a reality in which false, distorted or misleading information has become ubiquitous. In this context, we are inundated with a constant stream of news, claims and opinions, some of which may be false or manipulated for various purposes. The spread and accessibility of new technologies have amplified the problems of false information, which is an issue "almost as old as human history," as C.R. Sunstein acknowledges in his work entitled "Rumors, Gossip, and False Hearsay. How they spread, why we believe them, how we can defend ourselves," but which have now spread globally and on a large scale. This challenge has been fueled, at least in part, by the increasing role of Artificial Intelligence in the context of information and communication.

Indeed, over the past few years, AI has demonstrated that it can significantly influence social communication. On notable occasions, AI systems have directed political debate, influenced users' business choices and even altered public perceptions. Some of the best-known cases, such as the Brexit or 2016 U.S. elections, have raised concerns that automated systems known as "political bots" have helped distort public debate, often with the goal of matching human behavior. In addition to the concrete impact of such systems in specific events, there is an emerging trend of "computational propaganda," i.e., the tendency to engage in manipulative behavior, disseminated through social networks, with the aim of modulating public opinion, which is sometimes biased against the truth. Paradoxically, however, AI offers one of the most promising solutions for dealing with misinformation and fake news, as it turns out to be a technology defined as "selective" due to the capabilities described above. Much of the technology used to identify and manage misinformation is powered by AI, which allows journalists and "fact-checkers" to quickly verify large amounts of information, a task otherwise impossible without help from technology. However, it should be noted that social problems arising from humans, such as political polarization, lack of critical thinking and the distinction between truth and falsehood, cannot be completely solved by the use of AI.

7   The Concept of Transparency

In addition to the aspect of veracity of information, another fundamental value is that of transparency. It can be divided into an external dimension, which requires that users of digital technologies be informed about the use of Artificial Intelligence, and a more complex internal dimension, which requires that AI-based decisions in the communication and information sphere be "explainable" in their decision-making processes. To date, tools capable of examining the principles behind automated decisions in the online sphere have been lacking, giving rise to the "black-box" problem, or the difficulty of understanding how AI learning models work and what motivates them.

However, the recent European Regulation on Artificial Intelligence, presented by the European Commission in April 2021, has included in its general provisions "harmonized transparency rules for AI systems intended to interact with natural persons, emotion recognition systems, biometric categorization systems and AI systems used to generate or manipulate images or audio or video content," which could also have a significant impact in the information sphere. With this we understand the importance of the issue of transparency and the role it plays in addressing the issue of profiling and all its techniques. This term represents any form of data processing in an automated manner, processing aimed at "assessing personal aspects relating to a natural person, in particular for the purpose of analyzing or predicting aspects of that natural person's professional performance, economic situation, health, personal preferences, interests, reliability, behavior, location or movements," as stated in Article 4 of the Data Protection Regulation (GDPR).

Using Artificial Intelligence in relation to these techniques increases the possibility of obtaining much more detailed and personalized profiles of users using intelligent systems. In fact, when such individuals interact with these technologies, they reveal a set of personal preferences that the AI captures, analyzes and reworks thanks to the aforementioned deep learning and machine learning systems. In doing so, the systems are able to offer services and content that are increasingly personalized and that reflect users' tastes, satisfying different needs and demands. These profiling algorithms are used by most e-commerce platforms such as Amazon, Zalando, eBay etc. and also by the major search engines we know today, examples of Google, Firefox, Edge etc.

8   The Logic of the Algorithm

Some of the examples of these algorithms are Google's "Page Rank" and "RankBrain" or the "News Feed" service used by Facebook. Regarding the first case, Google's systems are applied according to two different categories of filters, one "traditional" and the other "intelligent." The traditional technique selects and sorts information according to its relevance, following objective parameters; the intelligent one, on the other hand, assigns greater or lesser relevance to different content, depending on the user's search. Since the end of 2015, this algorithm has been enhanced with the introduction of "RankBrain," a technique based on machine learning, which is not only capable of compiling a list of results based on relevance, but more importantly, capable of understanding the user's query without the input of keywords.

RankBrain's algorithm can, in addition, create common fingerprints as a result of analyzing connections between different user search queries, so that a more accurate answer can be received despite having used ambiguous terms. In the second case, Facebook's News Feed system, customizes content on each user's homepage based on their interests, creating an information bubble that reflects the user's tastes and opinions. This process leads people to be exposed primarily to information that confirms what they already believe, reinforcing their beliefs.

Although profiling may have advantages, such as receiving information consistent with personal preferences, it becomes problematic when it comes to access to objective and varied information. Such processes, therefore, become relevant in this area mainly due to two factors: the crisis of traditional information media and the spread of the so-called "networked information economy." In recent years, there has been a significant decrease in the use of traditional information media, as evidenced by the decline in newspaper sales in Italy. At the same time, social networks, such as Facebook, search engines and YouTube, are becoming primary sources of information for a growing number of people, especially among young people. While, as far as the "networked information economy" is concerned, we can say that it is characterized by decentralized production of information, where any user can contribute to the creation of news and content. Rapid and continuous access to information and communication tools is made possible by widely available electronic devices.

8.1   Global Information

This new information age is also characterized by open and global information flows, thanks to the development of digital infrastructure. It can therefore be said that, the use of Artificial Intelligence in the field of profiling has a double face, on the one hand such use results in multiple benefits, on the other hand, however, the use of these techniques could generate uncertain scenarios, in which the principles underlying the democratic system and the freedom to be informed are conditioned.

The integration of Artificial Intelligence into the content moderation process is a significant step forward in the digital world. However, this innovation raises critically important questions, especially regarding the possible presence of bias in the decision-making process. While algorithms are extraordinary in their inherent power, their effectiveness is profoundly affected by the quality and nature of the data used during the training phase. It is important to understand that although they are designed with the goal of analyzing and evaluating content objectively, algorithms can be shaped by the characteristics belonging to the training data. In cases where historical data reflect biases or inequalities that are already present in society, AI may unintentionally perpetuate and amplify these trends. For example, an algorithm might show a propensity to flag content from specific demographic groups more frequently or to treat certain topics or cultural languages unequally. To effectively address these possible biases, a careful and methodical approach must be taken by the platforms and developers responsible for implementing such algorithms.

Rigorous measures need to be developed to mitigate such distortions. This may require human intervention, the use of balanced and diverse data for training and maximum transparency in the implementation of moderation policies. Understanding and mitigating bias in moderation algorithms represents a crucial step towards creating a fair and inclusive online environment, where freedom of expression is protected without discrimination. The adoption of these advanced practices not only helps to ensure a fairer and more respectful virtual environment, but also represents an important demonstration of responsibility and commitment on the part of the platforms towards their users. This effort not only promotes user trust in the virtual environment, but constitutes a significant step towards building more inclusive and conscious online communities, in which diversity of thought is not only tolerated but appreciated as a fundamental resource for virtual environment. Enrichment of debate and global understanding. The use of algorithms in content moderation constitutes a step forward in managing online dynamics, but opens up a discussion on the incidence of possible bias. These algorithms, while powerful, are moldable by the quality and nature of the data used in their training. It should be underlined that, although designed for an objective evaluation of the contents, they can be affected by the peculiarities of the training data. If historical data reflects biases or inequities already present in society, AI can unintentionally perpetuate and amplify those trends. The most relevant consequence is the possible manifestation of a conservative behavior of the algorithms, which can classify certain types of content in an excessively restrictive way. This translates into a sort of "over-censorship", which, if not carefully monitored, risks limiting users' freedom of expression and precluding open discussion on certain topics. The implications of this over-censorship are far-reaching.

9 Some Perspectives

First, it can discourage users from actively participating in conversations, fearing that their expressions may be falsely labeled as inappropriate or offensive. This can lead to a decrease in the diversity of voices and points of view, going against the fundamental principle of freedom of expression. Furthermore, the excessive restriction of expressions can contribute to a sort of "homologation" of online discourse, preventing frank and open discussion on important issues and often creating controversies. In this way, the exchange of ideas and the formation of informed opinions can be compromised. It is therefore crucial to carefully monitor and, when necessary, correct the algorithms to avoid over-censorship. This can be achieved through a process of implementing user feedback mechanisms and adopting flexible moderation policies that take into account the context in which a given content is presented.

Ensuring correct implementation of moderation algorithms is essential to promoting an online environment that is inclusive, diverse and respectful of the freedom of expression of all users. This is a crucial commitment, which not only promotes user trust in the virtual environment, but constitutes a significant step towards building more inclusive and aware online communities. Diversity of thought is not only tolerated, but appreciated as a fundamental resource for enriching global debate and understanding.

Considering the risk of discrimination and bias within moderation algorithms represents an extremely relevant aspect in the digital landscape. These tools learn from historical data, and if this data reflects biases or inequalities already present in society, AI could unintentionally perpetuate and even amplify them.

The objective of guaranteeing fair and impartial moderation is thus transformed into a categorical imperative. Platforms and developers tasked with implementing these algorithms must take rigorous measures to mitigate bias. This commitment may involve establishing human review processes, using balanced and diverse data to train algorithms, and ensuring maximum transparency in the implementation of moderation policies. However, these measures, while important, represent only part of the solution. The interaction between Artificial Intelligence technologies and freedom of expression raises a number of crucial issues, including non-discrimination. This aspect is of particular importance, as AI can introduce risks of discrimination both in access to information and in the dissemination of the contents themselves.

A concrete example of this phenomenon is represented by AI-based recommendation algorithms, which can give rise to what is called an "echo chamber". In these situations, users are primarily exposed to content that reflects and reinforces their pre-existing opinions and beliefs. This process can have the effect of limiting the diversity of opinions and points of view, thus undermining information pluralism, which is a fundamental pillar of freedom of expression. Another critical element to consider is the tendency of Artificial Intelligence to inherit and amplify the biases and discriminations present in the data on which it is trained. In other words, AI may unintentionally favor certain groups or viewpoints over others if appropriate corrective measures are not implemented. These dynamics pose a significant challenge, requiring a thoughtful approach that ensures non-discrimination in the use of AI to disseminate information and promote freedom of expression.

The advancement of these intelligent technologies merges inexorably with the progressive decline of the central role of publishing in information, producing significant implications for the creation, distribution and use of information. This process has given rise to a complex evolution in the relationship between information pluralism, the opening of the publishing market and competition, creating an entanglement that is difficult to untangle. This synergy, closely connected to the selective and organizational function of AI, has had a significant impact on both the active and passive aspects of the right to information, addressing crucial issues in contemporary democracies. In the context of today's web-based information economy, we are witnessing a radically decentralized information production, with each user called upon to actively participate. This phenomenon is made possible by the widespread diffusion of electronic devices and the presence of open and global communication flows.

This wealth of networks, if carefully managed, has the potential to broaden the sphere of freedom and participation in democratic life, offering fertile ground for a richer and more nuanced public debate. However, when we focus on the passive profile of the right to information, we observe a change in the user's position. Although the production of information is distributed among various subjects and individuals, this information is still organized and made accessible by a small group of actors who hold the power to build bridges and connections across platforms, social networks and search engines. This limited set of actors, often with considerable economic power, have control over the volume and resonance of various voices in public space, giving rise to what can be described as potential "gatekeeping".

Search engines and social networks, as already widely anticipated, play a crucial role in organizing and providing information to users, they also make use of complex algorithms based on machine learning to personalize information based on browsing habits of individuals. In this scenario, the importance of education and information clearly emerges to ensure that each individual can contribute significantly to the public debate, maintaining a critical conscience at both an individual and collective level. However, the current landscape of information pluralism threatens this fundamental role, paving the way for what has been defined as a "bubble democracy".

10 Conclusions

The meeting point in the relationship between fundamental rights and new technologies in modern democratic contexts focuses on the essential principle of equality and non-discrimination. This is of particular relevance when considering the role of AI in selecting and moderating online content. It is important to note that AI, in its functioning, makes distinctions and choices based on the data at its disposal. This means that it can, if not adequately controlled, reflect and amplify the prejudices and stereotypes of its creators, and act according to processes that are difficult to scrutinize. This puts already disadvantaged groups in society at risk and can give rise to new forms of discrimination, requiring careful and intersectional analysis. AI-induced discrimination is considered insidious precisely because of its subtle nature. They are difficult to prevent, detect and remedy, as they are often rooted in the algorithms themselves, which are themselves complex and difficult to understand. Unfortunately, rather than promoting fairer and more impartial decisions, many instances of discrimination have emerged from systems based on deep learning and machine learning. Egregious cases have been documented, such as Amazon's personnel selection system that penalized women or the Compas system in the United States. In the latter case, a group of experts highlighted that the algorithm had a margin of error in predicting an individual's repetition of a crime, but what emerged as relevant was how the algorithm made errors. In fact, it systematically overestimated the risk for defendants of African American origin and underestimated the risk for whites. This poses a serious question of constitutionality when the algorithm is inherently biased due to the initial data feeding it. The importance of affirming the principle of non-discrimination in the context of new technologies and Artificial Intelligence has been underlined in various forums and documents, as in the case of the European Commission for the Efficiency of Justice (CEPEJ), an organization that carries out a key role in promoting fair and efficient justice in Europe. CEPEJ is actively involved in developing guidelines and recommendations to address the challenges posed by AI and emerging technologies in the legal sector. Furthermore, the Proposal for a European Union Regulation on artificial intelligence has referred to the implementation of this principle in several points, even if it only partially addresses the issue of data. Discriminatory processes have also been found in the fields of search engines and social media.

For example, Facebook was found to allow advertising targeting based on sensitive data, such as sexual preferences. The same goes for ProPublica, which demonstrated how some industries excluded groups by discriminating against them based on their race, such as Hispanic and black people. Another aspect concerns the visibility of contents within the platforms, with mechanisms such as the Shadow Ban which limit their diffusion based on non-explicit criteria. A much debated topic concerns online hate speech, which can be based on various factors such as sex, gender, sexual orientation, race, religion and disability. The role of major platforms such as Twitter, which has now become X, Instagram and Facebook in managing such behavior is the subject of controversy.

National laws and European Union regulations also try to address these problems, but the issue of limiting freedom of expression remains central. In this whole context, there is a risk that the algorithmic tools used to combat online hatred could end up silencing forms of expression that instead fall within the fundamental rights of information. It is a field that is still evolving and the balance between rights and duties remains an open challenge.

References

1. Balzola, A.: Edu-action. 70 tesi su come e perché cambiare i modellieducativi nell'era digitale. Meltemi, Milano (2021)
2. Boccia Artieri, G.: Stati di connessione. Pubblici, cittadini e consumatori nella (Social) Network Society. FrancoAngeli, Milano (2012)
3. Bonaiuti, G., Calvani, A. (ed.): Le tecnologie educative. Carocci, Milano (2017)
4. Couldry, N.: Sociologia dei nuovi media. Teoria sociale e pratiche mediali digitali. Pearson, Torino (2015)
5. Joosten, T.: The social media for educator. Jossey-Bass, London (2012)
6. Monga, M.: Turing. La nascita dell'intelligenza artificiale. Pelago, Milano (2021)
7. Ottolini, G., Rivoltella, P.C.: Il tunnel e il kayak. Teoria e metodo della Peer&Media Education. Franco Angeli, Milano (2014)
8. Patrut, M., Patrut, B. (ed.): Social Media in Higher Education: Teaching in Web 2.0. IGI Global, Hershey (2013)
9. Ranieri, M., Manca, S.: I social network nell'educazione: Basi teoriche, modelli applicativi e linee guida. Erickson, Trento (2017)
10. Silverstone, R.: Mediapolis. La responsabilità dei media nella civiltà globale. Vita e Pensiero, Milano (2009)
11. Varnelis, K.: Networked publics. MIT Press, Cambridge (2008)
12. Vittadini, N.: Social Media Studies. I social media alla soglia della maturità: storia, teorie e temi. FrancoAngeli, Milano (2018)

Inclusion: A Concept Too Young for Artificial Intelligence

Stefano Di Tore[1], Pio Alfredo Di Tore[2]([✉]), Umberto Bilotti[1], and Maurizio Sibilio[1]

[1] Università degli Studi di Salerno, Salerno, Italy
{sditore,ubilotti,msibilio}@unisa.it
[2] Università di Cassino e del Lazio Meridionale, Cassino, Italy
pioalfredo.ditore@unicas.it

Abstract. We attempted to prompt a GPT system to generate a response to a specific prompt and detail the steps the system followed to produce the generated response. For this purpose, we employed the open-source toolkit Gpt4all, utilizing the opensource language model LLama 7bit-quantized. The pretrained model was specialized on a series of thematic texts using the open-source libraries Langchain. The choice of toolkit, model, and libraries was determined by two factors: 1) the intention to utilize both fully open-source toolkits and models, and 2) the necessity to use toolkits and models capable of running on consumer-grade CPUs, within our available resources. This work focuses on the concepts of prediction, bias, and explainability that motivated the aforementioned experiment.

Keywords: AI · Explainability · Bias · Inclusion

1 Inclusion: A Concept Too Young for Artificial Intelligence

The virtually infinite customization possibilities associated with Artificial Intelligence have sparked immediate enthusiasm in the educational domain, particularly regarding the theme of inclusion. When queried broadly on this matter, ChatGPT responds with somewhat generic caveats:

"The relationship between education, inclusion, and artificial intelligence is a topic of increasing relevance in the digital era we inhabit. Artificial intelligence (AI) has the potential to revolutionize the education sector in multiple ways. Firstly, AI can personalize learning by adapting educational content to the individual needs of students, thereby enhancing the effectiveness of teaching. This inclusive approach is particularly beneficial for students with special needs, enabling them to receive tailored support.

However, it is crucial to address inclusion-related challenges when employing AI in education. Ensuring that access to AI-based educational technologies is equitably distributed is essential, so as not to leave any group of students behind due to their resources or socio-economic conditions. Furthermore, careful consideration must be given to privacy and data security issues when using AI-based systems to gather information about students.

Education plays a crucial role in ensuring that the integration of AI is inclusive and responsible. Teachers need to be trained to use these new technologies effectively and ethically, while educational policy-makers must develop regulations that protect students' rights and promote fair access to AI in education.

In summary, the relationship between education, inclusion, and artificial intelligence is complex but full of potential. If managed carefully and responsibly, the use of AI in education can contribute to creating a more inclusive and adaptable educational environment that meets the needs of all students, regardless of their differences" (ChatGPT 2023).

In reality, despite the easy enthusiasm and acknowledging the undeniable potential of artificial intelligence (not only generative artificial intelligence) in inclusive education, it seems appropriate to emphasize some fundamental issues that will likely systematically evade the "ethical" controls of artificial intelligence systems.

In this perspective, we need to briefly define some fundamental concepts in the field of AI, particularly anticipating the constructs of Bias and Explainability from the outset.

- Bias refers to systematic distortions or discriminations in the collection, analysis, or interpretation of data within the field of artificial intelligence. These biases can be introduced by training data or algorithm design and may lead to unfair or discriminatory outcomes.
- Explainability refers to the ability of an artificial intelligence model to explain in a comprehensible and transparent manner how it reached a specific decision or prediction. It is crucial for understanding how AI algorithms operate and for ensuring their accountability and ethics.

Regarding the risk of bias, it is essential to underscore how inclusion, as an educational principle, is still relatively young in its evolution, and the datasets used to train AI systems may reflect biases and past practices that do not align with the current conceptualization of inclusion. For instance, historical data used to train machine learning algorithms might contain discriminatory information regarding students with disabilities, students from ethnic minorities, or disadvantaged groups. These data can negatively influence AI decisions and recommendations in education, fostering disparities and discrimination instead of promoting an inclusive environment.

It is crucial to highlight that historical data is not inherently "wrong"; it genuinely reflects the society of the time in which it was collected. However, the critical point is that these data may contain biases, discriminations, or social practices that we now consider unacceptable or not in line with the current concept of inclusion.

Deleting or directly correcting them can be risky, as it might represent a form of historical revisionism or data manipulation. However, adopting a balanced approach is important. Instead of outright elimination or correction, organizations and researchers working with artificial intelligence must consider the possibility of mitigating bias through bias reduction techniques, data refinement, and the implementation of equity strategies in algorithm training. The mentioned techniques (from resampling to synthetic data generation to weight assignment) inherently carry a high risk of arbitrariness.

Furthermore, it is important to ensure transparency about the origins of the data and their potential limitations, enabling users to understand the historical contexts in which they were collected. This can contribute to creating a critical awareness of social and cultural evolution and promote reflection on how to improve inclusion without ignoring history. In any case, the need for transparency extends beyond the dataset on which the artificial intelligence system is trained or operates, also involving the algorithm's ability to explain how it reached a particular conclusion or decision. In this context, we refer to explainability, a key feature in AI that often finds itself in conflict with other desirable features such as speed or computational efficiency in market contexts.

Finally, and not insignificantly, we must consider the specific dimensions that have characterized the path of integration and inclusion in our country over the last fifty years. These have shaped an Italian situation very different from other European countries, not only in terms of regulations but also in the organization of the educational system itself. This unique situation makes it challenging to adopt tools designed and developed in different cultural contexts. In this regard, we refer to the introduction by Fabio Dovigo in the Italian edition of the Index for Inclusion.

2 Technical Tests of Transparency in Generative Artificial Intelligence

The recent attention to generative artificial intelligence systems has brought visibility, where needed, to the enduring categories of apocalyptics and integrators. In this text, we will attempt to steer clear of the Italian penchant for enlisting "regardless" in the ranks of Guelfs or Ghibellines. Instead, we will strive to identify some indispensable elements, in the opinion of the author, for exploring the conceptual perimeter defined by the topics of Artificial Intelligence, Machine Learning, Bias, and Explainability. This is not a theoretical-argumentative contribution or a literature review but rather the anticipation of a small experiment, in a broad sense, on explainability. We attempted to ask a generative artificial intelligence system to produce a response to a specific prompt and to detail the steps the system followed to generate the actual response. For this purpose, we used the open-source toolkit Gpt4all, with the open-source language model LLama 7bit-quantized. The pretrained model was specialized on a series of thematic texts using the open-source Langchain libraries.

The choice of toolkit, model, and libraries was determined by two factors:

1 - The desire to use both fully open-source toolkits and models.
2 - The need to use toolkits and models that could run on consumer-grade CPUs within our available resources.

Both factors are considered crucial to maintaining full control over the output. This contribution focuses on the concepts of prediction, bias, and explainability that motivated the aforementioned experiment.

3 Prediction and Prejudice

All generative artificial intelligence systems are built on powerful forms of machine learning where algorithms "learn" to predict specific outcomes from patterns and structures in vast datasets. Essentially, this involves a form of forecasting and imitation.

Prediction is a crucial concept in Generative AI, as it enables systems to generate new data that resembles the data on which they were trained. This can be valuable for various applications, including the creation of new creative content, problem-solving, and forecasting future events.

Several methods of prediction exist in Generative AI. One of the most common methods is using neural networks. Neural networks are a type of machine learning algorithm that can be employed to learn relationships between inputs and outputs. When a neural network is trained on a dataset, it can generate new data that is similar to the training data.

Another method of prediction in Generative AI involves the use of optimization techniques. Optimization techniques can be employed to find parameter values for a model that generate data most similar to the training data.

Prediction in Generative AI is used in real-world applications such as:

- Creative Content Generation: AI systems can predict and generate new, diverse content, ranging from art and music to writing, by learning patterns from existing creative works.
- Problem Solving: Generative AI can be applied to predict potential solutions to complex problems by analyzing patterns in relevant datasets, aiding in decision-making processes.
- Future Event Forecasting: AI models can predict future events based on historical data, helping in areas like financial forecasting, weather prediction, and trend analysis.

It is important to note that while these applications showcase the potential of Generative AI, the underlying algorithms must be carefully scrutinized for biases, and the explainability of generated outputs is crucial for maintaining transparency and trust in the system.

The final outcome of this prediction exercise is clearly tied to the datasets on which the system operates and those on which it has been trained. If the data is biased, the system will also be biased. For instance, when probation committees in the United States began using data to predict the risk of recidivism, they confronted a century of racism embedded in the data. Stories of bias in the American criminal justice system are reflected in the data used to train machine learning algorithms, which can then reproduce and amplify those patterns of injustice.

The example of the American criminal justice system is paradigmatic, and various studies address the issue, focusing on specific aspects:

An experiment conducted by ProPublica in 2016 revealed that risk prediction software used in many American courts was heavily biased against African Americans. The experiment found that the software was more likely to predict that African Americans would reoffend, even when there was no evidence to support this claim. This led to a disproportionate number of African Americans being incarcerated (Angwin, J., Larson, J., Mat-tu, S., & Kirchner, L. (2016, May 23). Machine bias in criminal justice. ProPublica).

Another experiment conducted by MIT in 2017 found that facial recognition systems are more likely to misidentify people of color as suspects. The experiment discovered that facial recognition systems were 35% more likely to misidentify African Americans as suspects compared to whites. This resulted in a disproportionate number of people of color being falsely arrested (Buolamwini, J., & Gebru, T. (2018, February 1.). Gender shades: Intersectional accuracy disparities in commercial facial analysis. Proceedings of the ACM on Human-Computer Interaction, 2(1), Article 14).

An experiment conducted by the New York City Department of Correction in 2018 found that the prison cell assignment system was heavily biased against African Americans. The experiment revealed that African Americans were more likely to be assigned to overcrowded and unsanitary cells compared to whites. This resulted in a disproportionate number of African Americans contracting illnesses while in prison (Peck, J. 2018, November 13. New York City jails have a racial bias problem. The New York Times).

What makes predictive tools interesting for analysis in the educational context is the need to decide how to approach historical injustice when creating such tools. If one seeks to maintain a neutral stance by merely building a more accurate tool, there is a risk of reproducing and solidifying the underlying patterns of injustice.

The theme has long been recognized in literature. For a systematic review, refer to (Danaher 2019) and (Alan Turing Institute 2019).

4 Explainability

Based on the above, it is evident that the possibility of free access to datasets and language models used in generative artificial intelligence systems is a fundamental condition for democracy, ultimately making it crucial in an educational context. However, it is important to note that free access to models, datasets, and code does not entirely solve the problem. Many AI algorithms operate as "black boxes," not explaining how they reach predictions or decisions.

In summary, a Black Box is a system whose internal workings are unknown or not easily understandable. In the field of AI, Machine Learning models are often considered black boxes because their internal workings are often highly complex and difficult to comprehend.

Several reasons contribute to Machine Learning models functioning as Black Boxes. Firstly, ML models are often trained on vast sets of input and output data. This can make it challenging (and resource-intensive) to explain to a human how the model learned the relationships between inputs and outputs. Secondly, ML models often rely on neural networks, which are highly complex mathematical systems. Neural networks can learn non-linear relationships between inputs and outputs, but explaining how these relationships are learned is difficult.

The fact that ML models are black boxes can pose issues in various situations. Firstly, it can make it challenging to understand how the model arrived at a particular decision, which can be problematic in critical applications like autonomous driving systems or medical diagnosis assistance. Secondly, the opacity of ML models can hinder the identification and resolution of biases in the models. This presents a clear problem when the model is used in an application where fair and unbiased decision-making is crucial.

The ability to clearly and comprehensibly explain the decision-making process of AI algorithms, providing a rationale or justification for their predictions or actions, is defined as explainability. The scientific literature on the concept of explainability is abundant, and a notable contribution to the field of model interpretability is the paper by Lipton (Lipton, Zachary C. "The mythos of model interpretability." arXiv preprint arXiv:1606.06565 (2016). This paper clarified the concept of interpretability, identified challenges and trade-offs associated with model interpretability, and stimulated further research in the field. It discusses the notion of explainability in ML models, arguing that there is no single definition of explainability and that the goals of explainability can vary depending on the context.

The paper identifies three distinct notions of explainability:

- Local Explainability: The ability to understand the decisions made by the model for a single data point.
- Global Explainability: The ability to understand the generalizations made by the model over a dataset.
- Causal Explainability: The ability to understand the causal mechanisms the model is learning.
- The author argues that none of these notions of explainability is sufficient alone and that a multi-perspective approach is necessary for the explainability of ML models.

The paper concludes by discussing some of the challenges and trade-offs associated with the explainability of ML models. The author asserts the importance of finding a balance between explainability and accuracy, considering the needs of different stakeholders when designing and implementing ML systems.

The importance of explainability lies in several factors. Firstly, explainability promotes transparency and accountability of AI algorithms. For instance, in sectors such as healthcare and justice, where decisions can significantly impact people's lives, understanding why an algorithm suggested a particular diagnosis or rendered a specific judgment is crucial. The ability to explain the reasons for AI decisions allows verification of their reliability and ensures they are free from discrimination or bias.

Secondly, explainability contributes to building trust and social acceptability towards AI systems. Algorithms operating as "black boxes" can evoke concerns and distrust from users and the general public. Explaining how an algorithm reaches its predictions or decisions helps dispel doubts and provides an intelligible justification for its actions, thereby increasing trust in the system.

5 Technical Tests of Explainability

As anticipated, we attempted to solicit a response from a generative artificial intelligence system by providing a specific prompt and detailing the steps the system followed to generate the actual response. For this purpose, we utilized the open-source toolkit Gpt4all, employing the open-source language model LLama 7bit-quantized. The pretrained model was specialized on a set of thematic texts using the open-source libraries Langchain.

Specifically, we presented the language model with a semantic context, a text document to serve as the framework for generating the response. In our case, the context was the Wikipedia page dedicated to the entry "Pedagogy."

GPT4All constitutes an ecosystem for training and deploying large, customized language models that operate locally on consumer-grade CPUs.

The process is relatively straightforward and unfolds through the following steps:

1. Load the GPT4All model.
2. Utilize Langchain to retrieve and load our documents.
3. Partition the documents into small, manageable blocks for the Embeddings.
4. Employ the FAISS libraries to create a vector database with the embeddings.
5. Conduct a similarity search (semantic search) on the vector database based on the question we intend to pass to GPT4All: this will be used as the context for our question.
6. Formulate the question and provide the context to GPT4All using Langchain; await the response.

An Embedding is a numerical representation of a piece of information, such as text, documents, images, audio, etc. The representation captures the semantic meaning of what is embedded.

From the image, we can deduce the text chunks extracted from the PDF file used as context that the system deemed relevant for generating the response to our prompt ("What is pedagogy?"). It is worth noting that the response, unless otherwise specified, is in English, even though the context was provided in Italian.

6 Conclusions

The theme of explainability is closely intertwined with that of transparency and the feasibility of open access to systems. While, in theory, explainability may appear to be a paramount objective for all language models, in a market context, it can conflict with other aspects of AI, such as accuracy or computational efficiency. More interpretable models might sacrifice a certain amount of accuracy or require more computational resources compared to more complex models. Therefore, in balancing the need to explain AI decisions with the demand for efficient performance, different contexts may lean towards performance at the expense of transparency.

This clearly poses the challenge of establishing a set of non-generic rules applicable universally across international contexts.

References

Alan Turing Institute. The ethics of artificial intelligence. Alan Turing Institute (2019)

Danaher, J.: Bias in artificial intelligence: a modern approach. Oxford University Press (2019)

Johnson, R.L., et al.: The Ghost in the Machine has an American accent: value conflict in GPT-3. arXiv preprint arXiv:2203.07785 (2022)

Linardatos, P., Papastefanopoulos, V., Kotsiantis, S.: Explainable AI: a review of machine learning interpretability methods. Entropy **23**(1), 18 (2020). MDPI AG. Retrieved from https://doi.org/10.3390/e23010018

Non-player Character Smart in Virtual Learning Environment: Empowering Education Through Artificial Intelligence

L. Campitiello[(✉)], V. Beatini, and S. Di Tore

University of Salerno, Fisciano, Italy
lcampitiello@unisa.it

Abstract. In recent years, the evolution of intelligent NPC (Non-player Characters) generated by artificial intelligence (AI) has revolutionized numerous sectors, from entertainment to education and virtual assistance. Intelligent NPC, delineated as virtual entities within video games and interactive simulations, exhibit a notable proficiency in artificial intelligence, autonomy, interaction, and decision-making. These entities, portrayed as realistic digital representations, assume a pivotal role in enhancing the user experience by dynamically adapting to the choices made by players. In the educational sector, NPC can improve learning and interaction between students, for example they can be used as virtual tutors to assist students and answer questions to provide explanations for complex concepts. Therefore, this work aims to develop an intelligent NPC to be integrated within the Inclusive Virtual Museum to promote knowledge of cultural heritage in the students.

Keywords: Artificial Intelligence · NPC smart · Virtual Reality

1 Introduction

In the 1950s, the concept of artificial intelligence was introduced as a mere theory that reflected the manifestation of human intelligence by machines [1]. In the contemporary era, marked by rapid technological progress, artificial intelligence has transcended the theoretical phase to become a tangible application [2], rooted in various aspects of society, through its widespread presence in various personal electronic devices.

The field of artificial intelligence (AI) has roots that date back to different eras, but the current field of study took shape in the 20th century. One of the pioneers in the theorizing of a machine capable of thinking intelligently was the mathematician Alan Turing, who in 1950 published the article "Computing Machinery and Intelligence". In this article Turing proposed the famous "Turing test" or "human behavior imitation test", which aims to evaluate whether a machine can demonstrate intelligent behavior indistinguishable from human behavior. Turing's work constitutes one of the key points in the evolution of artificial intelligence (AI), but other pioneers have contributed significantly, including John McCarthy, Marvin Minsky, Nathaniel Rochester and Claude Shannon, who in 1956 they organized the Dartmouth Conference, an event considered the official birth of the field of AI.

The field of study of Artificial Intelligence (AI) mainly focuses on the development of systems and programs capable of emulating the cognitive processes of human intelligence, such as learning, reasoning, problem solving and perception. This constantly evolving sector has significant potential to revolutionize many fields, including education, medicine, industry and other sectors. Some applications of artificial intelligence include: 1) machine learning, which involves training computers to learn from data in order to improve performance; 2) artificial neural networks, which are computational models that emulate the configuration of the brain and find application in resolving intricate issues; 3) natural language processing (NLP), i.e. the competence of computers in understanding and producing human language; 4) artificial vision, which implies the machine's ability to interpret the visual world; 5) robotics, which focuses on the ability of robots to perform complex actions in the physical environment; 6) reasoning and decision making, which aim to provide support for decision making by analyzing data to formulate predictions.

In particular, machine learning (ML) constitutes a subset of artificial intelligence designed to replicate the experiential learning inherent in human intelligence. Through the use of computational algorithms, machine learning effectively learns and improves analytical capabilities [1]. These algorithms leverage datasets that include input and output information to identify patterns, enabling autonomous learning and decision making. This iterative process allows the machine to receive inputs, predict outcomes, and more sophisticatedly, "deep neural networks" use hierarchical layers to manage the final output. Similar to the functionality of the human brain, the machine establishes "neuronal" connections similar to "dendritic" connections across various levels of hierarchical data. This innovation has given rise to a new form of artificial intelligence known as "deep learning" [3]. Natural language processing (NLP), on the other hand, focuses on enabling computers to understand, interpret and generate human natural language, facilitating the use of natural language for communication between machines [4]. NPL finds application in machine translation, chatbots, sentiment analysis and other related fields.

The emerging field of Educational Artificial Intelligence (EAI) represents a fusion of artificial intelligence and learning sciences, seeking to improve education through adaptive, flexible and customizable learning environments. Furthermore, EAI uses accurate calculations to articulate knowledge that may present ambiguity in the human sciences [4]. The applications of artificial intelligence in education span various dimensions: 1) knowledge representation, this involves the use of a data structure that computers can accept to articulate knowledge; 2) machine learning, enabling computers to discern their own decision-making processes through practical applications, using statistical methods to ultimately solve problems [5]; 3) Deep learning, an efficient feature extraction method, capable of abstracting more complex features from data to obtain a more essential characterization; 4) natural language processing, an attempt to allow computers to understand human natural language, facilitating its use in communicating with computers; 5) intelligent agent, a program with the ability to autonomously complete a series of operations in active service mode; 6) affective processing, the process through which humans program machines to recognize, understand, process and simulate human emotions [4].

In education, artificial intelligence (AI) can offer significant potential to revolutionize the way students learn and teachers teach. AI can be used to create personalized educational programs for students, adapting educational materials to their needs and analyzing data on student performance and preferences [6]. Furthermore, taking advantage of artificial intelligence intelligent NPC (Non-player characters) can be created that can provide assistance, answer questions, explain concepts and offer support to students during learning activities. Using NPC in education can make learning more engaging and interactive, giving students opportunities to experiment and learn in a virtual environment.

2 Non-player Character Smart in Virtual Learning Environment

Intelligent NPC (Non-Player Characters) fall within the scope of Artificial Intelligence (AI) and can involve different AI sub-disciplines, such as Machine Learning (ML) and Natural Language Processing (NLP), depending on their specific characteristics and functionality. Intelligent NPC can benefit from machine learning to improve their ability to adapt to the game environment and learn from player behaviors. For example, using machine learning algorithms, NPC can adapt to player tactics, learn strategies and improve their performance over time. When NPC are designed to communicate with players through natural language, then NLP becomes relevant. NPC can have more realistic and responsive conversations with players, recognizing written or spoken language and responding coherently. NPC interacting in a virtual environment that includes visual elements, such as 3D games or virtual reality, can use computer vision to perceive the surrounding world, navigate complex environments, or recognize objects and characters. If NPC interact with users through audio, they can use sound processing to recognize voice commands or react to sounds in the game environment.

Artificial intelligence (AI) plays an essential role in the evolution of NPC, allowing them to act in a more autonomous and sophisticated way. Unlike traditional NPC, who may exhibit predefined or limited behaviors, intelligent NPC are designed to be more adaptable and capable of making decisions on their own. These virtual characters, if properly developed, can help make the game more engaging and realistic.

In videogames, AI is used to create sophisticated behaviors for NPC, allowing them to make decisions based on context, learn from player behaviors, and emulate human intelligence in various forms. Applications of AI in videogame NPC can include strategy simulation, machine learning to improve NPC performance, natural language processing to make conversations with NPC more realistic, and much more. The goal is to offer players more engaging and realistic gaming experiences thanks to the interaction with NPC who appear to behave intelligently and reactively in the virtual environment.

Furthermore, a distinction must be made between NPC and avatars. Avatars represent a character controlled directly by a human player in a video game or virtual environment. Avatars are used to interact with the virtual world, perform actions and make decisions. In other words, avatars are an extension of the player in the virtual world and can be customized in various ways.

NPC, on the other hand, are virtual characters within a video game or virtual environment that are not directly controlled by human players but are programmed to act autonomously and play different roles, for example they can be opponents, allies or neutral characters. NPC are used to populate the game's virtual world and interact with players. In short, NPC are virtual characters that act autonomously in the game and are not controlled by the players, while avatars are the representations of the players themselves, through which they control the actions within the game. Both are important elements in video games, but they have different roles and are handled differently within the gaming experience.

The presence of intelligent NPC (Non-Player Characters) in virtual reality represents a significant element that contributes to enriching the user experience. In virtual reality, intelligent NPC can be programmed to act autonomously, adapting to users' decisions and actions, providing a higher level of realism and immersion. For example, in educational games or simulations, intelligent NPC could offer more complex dialogues, respond to perceived emotions in users or adapt their behavior based on users' choices. The introduction of intelligent NPC into virtual reality can find practical applications in virtual therapy or education. Indeed, in recent years virtual reality (VR) technology has been considered as an effective tool to facilitate the teaching-to-learning process. The technical characteristics of VR make it different from other ICT (Information and Communication Technology) applications as it is able to generate a digital environment that simulates real or imaginary environments by involving the human senses, such as sight, hearing and touch.

Virtual reality systems possess characteristics related to immersion, interaction, imagination and intelligence, and can find application in education through: 1) the creation of virtual environments (VE); 2) multi-sensory channels for user interaction; 3) the user's immersion in the VE; 4) intuitive interaction through natural manipulation in real time [4]. Virtual Learning Environments (VLE) can be defined as VE based on a pedagogical model that contains one or more pedagogical objectives, providing users with experiences that do not they can experience in the real world and contributes to specific learning outcomes. 3D VLE are social network learning environments that provide 3D simulations or learning objects (which cannot be obtained in the real world) to support teachers and students in the teaching-learning process. The theory of situational cognition underscores the significance of the social context in knowledge construction, and 3D VLE play a pivotal role in facilitating the creation of social situations and relationships. This, in turn, supports students in experiential learning [4]. Through experiential learning, students can engage in a given situation, fostering active thinking and experimentation. While real-world experiences are valuable, physical constraints and environmental limitations may hinder students from accessing them consistently. Simulated experiences using digital media offer a comparable alternative to real-world scenarios, offering considerable flexibility in their construction. Consequently, students can actively participate in activities, receiving realistic information and perceptible sensory stimuli. VLE provide a unique opportunity for individuals to transcend bodily constraints and prior experiences, enabling them to undergo various state changes and immerse themselves in an environment that closely approximates the real world.

3   Development of a Smart NPC for the Inclusive Virtual Museum

In Virtual Learning Envoirments, NLP technology is widely used to train learning agents who can engage in conversations with students. Students, using NLP technology, can converse with intelligent NPC, and in some cases, this interaction can help reduce anxiety levels. Machine learning and deep learning technologies are also applied in the creation of VLE systems to build a database and monitor the learning process in students in order to implement personalized teaching [7]. Using VLE, teachers can analyse students' attitudes and create personalized teaching in order to improve teaching efficiency. Another application of artificial intelligence is affective computing, which can detect students' behaviors and emotional states. Affective computing technology has often proven to be particularly effective for detecting the emotional states of children with autism in the virtual environment, demonstrating that the latter show fewer negative emotions than in reality [8]. In the educational sector, virtual environments can help improve students' academic performance and motivation, since the key to success lies in the use of virtual agents capable of interacting with students and offering interactive guidance.

Based on these considerations, an intelligent NPC was designed to be implemented within the Inclusive Virtual Museum [9] to interact with students and promote knowledge of cultural heritage. The inclusive virtual museum was developed through the *Unity 3D* graphics engine and inside it a room was created in which the digital heritage of the archaeological finds scanned at some Campania museums was inserted, such as the National Archaeological Museum of Sannio Caudino in Benevento, the Archaeological Museum of Carife, the De Chiara De Maio Foundation and the Filangieri Museum of Naples. The Virtual Museum was created taking into account the principles of Universal Design for All (UDL) in order to maximize accessibility by providing multiple ways of interacting, expressing and representing information. In this way it is possible to meet the different needs of students and offer an effective learning environment for everyone. In fact, in the virtual museum it is possible to modify some parameters, such as lighting, the height of the avatar, the fonts in the panels in order to adapt the virtual environment for students with special educational needs and guarantee a satisfactory experience.

To ensure an engaging experience and greater interaction in the museum room, an intelligent NPC capable of dynamically interacting with students and answering their questions has been implemented. In particular, the face of the NPC was created using the 3D scan of the face of a researcher from our team and then the Character Creator software was used to generate it. The NPC was trained to answer questions related to Samnite culture and to provide further information to students regarding the archaeological finds placed in the museum room. In this way the NPC can adapt to students' needs and offer personalized learning, as well as providing support on complex topics. Furthermore, interacting with the NPC can help students develop communication and social skills, as well as making learning more fun and motivating. In summary, it is possible to argue that the addition of an intelligent NPC can improve the students' learning experience, making the virtual environment more engaging and personalized (Figs. 1 and 2).

Fig. 1. NPC smart within the inclusive virtual museum

Fig. 2. NPC smart that provides support to the user by answering questions

4 Conclusion

Avatars and intelligent NPC are revolutionizing the virtual environment in ways that will profoundly change our interaction with the digital world. Thanks to artificial intelligence it is possible to create systems capable of learning and adapting, providing solutions to complex problems and tasks that require human intelligence. AI is a continually advancing field with the potential to revolutionize a broad spectrum of sectors, encompassing medicine, industry, education, and various others. The continuous evolution of artificial intelligence and the growth of data processing will allow the creation of increasingly convincing and interactive intelligent avatars and NPC. This will raise further questions about digital identity, privacy and security.

In the educational sector, the creation of intelligent NPC in virtual learning environments can promote the development of the teaching-learning process, adapting to student needs and offering personalized learning. NPC can also provide support on complex topics and interact with students to make learning more fun and motivating.

References

1. Bini, S.A.: Artificial intelligence, machine learning, deep learning, and cognitive computing: what do these terms mean and how will they impact health care? J. Arthroplast. **33**(8), 2358–2361 (2018). https://doi.org/10.1016/j.arth.2018.02.067
2. Topol, E.J.: High-performance medicine: the convergence of human and artificial intelligence. Nat. Med. https://doi.org/10.1038/s41591-018-0300-7
3. Haeberle, H.S., Helm, J.M., Navarro, S.M., et al.: Artificial intelligence and machine learning in lower extremity arthroplasty: a review. J. Arthroplast. (2019). https://doi.org/10.1016/j.arth.2019.05.055T
4. Wei, X., Jia, H.: A review of the application of artificial intelligence in the virtual learning environment. In: 2021 Tenth International Conference of Educational Innovation through Technology (EITT), pp. 79–82. IEEE. (2021)
5. Bundy A. Preparing for the future of artificial intelligence[J]. 2017
6. Rus, V., D'Mello, S., Hu, X., Graesser, A.: Recent advances in conversational intelligent tutoring systems. AI Mag. **34**(3), 42–54 (2013)
7. Yao, F., Zhang, C., Chen, W.: Smart talking robot Xiaotu: participatory library service based on artificial intelligence. Library Hi Tech (2015)
8. Marchi, E., Schuller, B., Baird, A., et al.: The ASC-inclusion perceptual serious gaming platform for autistic children. IEEE Trans. Games **11**(4), 328–339 (2018)
9. Campitiello, L., Caldarelli, A., Todino, M.D., Di Tore, P.A., Di Tore, S., Lecce, A.. Maximising accessibility in museum education through virtual reality: an inclusive perspective. Italian J. Health Educ. Sport Inclusive Didactics **6**(4) (2022)

Author Index

F. Palomba and C. Gravino (Eds.): WAILS 2024, LNCS 14545, p. 139, 2024.
https://doi.org/10.1007/978-3-031-57402-3